Klaus Kröncke

Stability of Einstein Manifolds

Klaus Kröncke

Stability of Einstein Manifolds

Südwestdeutscher Verlag für Hochschulschriften

Impressum / Imprint
Bibliografische Information der Deutschen Nationalbibliothek: Die Deutsche Nationalbibliothek verzeichnet diese Publikation in der Deutschen Nationalbibliografie; detaillierte bibliografische Daten sind im Internet über http://dnb.d-nb.de abrufbar.

Bibliographic information published by the Deutsche Nationalbibliothek: The Deutsche Nationalbibliothek lists this publication in the Deutsche Nationalbibliografie; detailed bibliographic data are available in the Internet at http://dnb.d-nb.de.

Coverbild / Cover image: www.ingimage.com

Verlag / Publisher:
Südwestdeutscher Verlag für Hochschulschriften
ist ein Imprint der / is a trademark of
OmniScriptum GmbH & Co. KG
Heinrich-Böcking-Str. 6-8, 66121 Saarbrücken, Deutschland / Germany
Email: info@svh-verlag.de

Herstellung: siehe letzte Seite /
Printed at: see last page
ISBN: 978-3-8381-3908-1

Zugl. / Approved by: Potsdam, UP, Diss., 2013

Contents

Chapter 0

Introduction

The Einstein-Hilbert functional associates to each metric the integral of its scalar curvature. It is a natural geometric functional because it can be considered as the mean over all curvatures on a given Riemannian manifold. It first appeared in the context of general relativity ([Hil15]) since the famous Einstein equations arise as the Euler-Lagrange equations of this functional. It is also of great interest in geometry. The famous Yamabe Problem was resolved by solving the Euler-Lagrange equation of the Einstein-Hilbert functional restricted to the conformal class of a metric (see [Sch84]).

The Ricci flow is a geometric flow first introduced by R. Hamilton in [Ham82]. It is a kind of nonliner heat equation for metrics that tends to smooth out irregularities in the metric. For any metric on a given compact surface, the volume-normalized variant of the Ricci flow starting at the metric converges to a metric of constant Gaussian curvature. In higher dimensions, the Ricci flow is much less understood and there are many open problems. On the other hand, Ricci flow techniques helped to open famous problems from geometry, e.g. the famous Poincaré conjecture ([Per02; Per03]) and the differentiable sphere theorem ([BS09]).

This thesis studies the Einstein-Hilbert functional and its variation at Einstein metrics and the Ricci flow close to Einstein metrics. It can be divided into two parts where the first part consists of Chapters 2,3 and 4 and the second part Chapters 5 and 6. Throughout, we deal with compact manifolds.

In Chapter 2, we introduce the Einstein-Hilbert functional and we summarize some well-known facts about its variational theory. The critical points of the Einstein-Hilbert action, when restricted to metrics of unit-volume, are precisely the Einstein metrics and they are always saddle points. The second variation admits a contrasting variational behavior in different directions of changes of the metric.

Einstein metrics are always local (even global) minima of the Einstein-Hilbert action restricted to unit-volume metrics in their conformal class. In contrast, the second variation restricted to the tangent space of the manifold $\mathcal{C}_1$ of unit-volume metrics with constant scalar curvature has finite coindex. We call an Einstein manifold stable if the second variation of the Einstein-Hilbert action is nonpositive on $T_g\mathcal{C}_1$. If this is not the case, we call the manifold unstable.

More precisely, the tangent space of $\mathcal{C}_1$ splits as

$$T_g\mathcal{C}_1 = T_g(g \cdot \mathrm{Diff}(M)) \oplus \mathrm{tr}_g^{-1}(0) \cap \delta_g^{-1}(0),$$

and because the Einstein-Hilbert functional is a Riemannian functional, its second variation clearly vanishes at $T_g(g\cdot \mathrm{Diff}(M))$. On $\mathrm{tr}^{-1}(0)\cap\delta^{-1}(0)$, it is given by $-\frac{1}{2}\Delta_E$ and Δ_E is a Laplace-type operator called the Einstein operator. The elements in its kernel are called infinitesimal Einstein deformations because they correspond to non-trivial curves of Einstein metrics through g.

The Einstein operator (or equivalently, the Lichnerowicz Laplacian) also appears on various occasions in physics. Solutions of $\Delta_E = 0$ are gravitational waves in the Lorentzian case (see [FH05]). Properties of the Einstein operator also play a role in the stability of higher-dimensional black holes ([GH02; GHP03]) which appear in higher-dimensional gravity theories.

In Chapter 3, we study concrete examples. First, we mention some well-known stable and unstable Einstein manifolds in Section 3.1. After that, we study flat compact manifolds and compute the dimension of the space of infinitesimal Einstein deformations explicitly in terms of the holonomy (Proposition 3.2.4). Then we discuss the stability properties of products of Einstein spaces and compute the index and the nullity of the quadratic form $h \mapsto (\Delta_E h, h)_{L^2}$ on the product in terms of mulitiplicities of certain eigenvalues of the Laplace-Beltrami operator on the factors and of the index and the nullity of this quadratic form on the factors (Proposition 3.3.7).

Chapter 4 is devoted to the study of stability under certain curvature assumtions. We first mention some results by N. Koiso which imply stablity under sectional curvature bounds. We then extend these results slightly (Propositions 4.2.2 and 4.2.5). Some eigenvalue bounds on the Einstein operator under curvature assumptions are given in Propositions 4.2.7 and 4.2.9.

We also prove some stablity criteria involving a quantity written in terms of the Weyl tensor (Theorems 4.3.4 and 4.3.7). Using an explicit expression of the Gauss-Bonnet formula for six-dimensional Einstein manifolds, this allows us to prove a stability criterion in dimension six which involves the Euler characteristic of the manifold (Theorem 4.5.4). Similarly to our considerations involving the Weyl tensor, we can prove stablity criterions for Kähler-Einstein manifolds involving the Bochner tensor (Theorems 4.6.7 and 4.6.8).

The proofs are based on the two Bochner formulas (4.2) and (4.3) for the Einstein operator and on estimates of the curvature action $\mathring{R}$ on symmetric $(0, 2)$-tensors.

In the second part of the work, we consider the Ricci flow close to Einstein metrics. We say that an Einstein metric (M, g) is dynamically stable if any Ricci flow (in an appropriate form) starting close enough to an Einstein metric converges (perhaps after pulling back by a 1-parameter family of diffeomorphisms) to an Einstein metric close to (M, g). Furthermore, we call (M, g) dynamically unstable if there is an ancient solution of the Ricci flow, which converges (perhaps modulo diffeomorphism) to (M, g) as $t \to -\infty$.

We build upon results from [GIK02; Ses06; Has12; HM13] for Ricci-flat metrics and [Ye93] for Einstein manifolds. The interesting point is the following: Although the Ricci flow is not the gradient flow of the Einstein-Hilbert functional, its behavior close to an Einstein metric is strongly related to the behaviour of the Einstein-Hilbert functional close to it.

With the use of the λ-functional and its variational theory, stability and instability assertions for compact Ricci-flat manifolds were proven in [Has12; HM13]. We transfer these results to non Ricci-flat Einstein metrics and use similar methods to those in [Has12; HM13].

We study negative Einstein metrics in Chapter 5 and positive Einstein metrics in Chapter 6. In both cases, the strategy is essentially the same. We introduce the functionals μ_+ and ν_- on the space of metrics which are nondecreasing under the Ricci flow variants (5.5), (6.1), respectively. We consider their well-known second variation formulas at Einstein metrics. Considering simplified expressions of these formulas, we see that they are of a similar nature to the second variation of the Einstein-Hilbert functional.

From these, it is easy to see that dynamical stability implies stablity with respect to the Einstein-Hilbert functional. The converse direction is much harder to prove and it is only true under additional assumptions.

In both chapters, we first prove stability/instability results which rely on the additional assumption that all infinitesimal Einstein deformations are integrable (Theorems 5.4.13, 5.4.14 and Theorems 6.4.7, 6.4.8, respectively). In the positive case, we also assume that $2\mu \notin \mathrm{spec}(\Delta)$ where μ is the Einstein constant. We obtain dynamical stability of an Einstein manifold (M, g) if it is stable with respect to the Einstein-Hilbert functional and if the smallest nonzero eigenvalue of the Laplacian satisfies $\lambda > 2\mu$. The convergence speed is exponential and we do not have to pull the flow back by diffeomorphisms. Dynamical instability holds if one of the two conditions fails.

Then we prove stability/instability results without the integrability condition and without the assumption $2\mu \notin \mathrm{spec}(\Delta)$ (Theorems 5.5.5, 5.5.6 and Theorems 6.5.8, 6.5.9, respectively). We then obtain dynamical stability if we assume that (M, g) is a local maximum of the Yamabe functional and that the smallest nonzero eigenvalue of the Laplacian satisfies $\lambda > 2\mu$. The convergence speed is polynomial and the convergence is modulo diffeomorphism. We have dynamical instability if (M, g) is not a local maximum of the Yamabe functional or $\lambda < 2\mu$.

The central tools are Lojasiewicz-Simon inequalities for the functionals μ_+ and ν_-. Another importent step is to prove local maximality of the functionals under the stability conditions mentioned above. In the integrable case, we furthermore prove transversality estimates which ensure that we do not have to pull back the flow by diffeomorphisms. The proofs of these three important properties mostly rely on Taylor expansion and careful estimates of the error terms. In the nonintegrable case, we apply a general Lojasiewicz-Simon inequality proven in [CM12].

From the previous results, it is not clear what to expect when the Einstein manifold is a local maximum of the Yamabe functional and the smallest nonzero eigenvalue of the Laplacian is exactly 2μ. We give a partial answer to this question in Section 6.6 and prove dynamical instability of $\mathbb{C}P^n$ (Theorem 6.6.3).

Chapter 1

Mathematical Preliminaries

In this short chapter, we recall some definitions and identities, fix sign conventions for the Riemann curvature tensor and the Laplacian and fix some notation. Throughout this thesis, any manifold is smooth, compact and connected and its dimension is at least 3 (unless the contrary is explicitly asserted).

Let M^n be a manifold and g be a Riemannian metric on it. We define the Riemann curvature tensor (as a $(1,3)$-tensor) with the sign convention such that

$$R_{X,Y}Z = \nabla_X \nabla_Y Z - \nabla_Y \nabla_X Z - \nabla_{[X,Y]} Z.$$

As a $(0,4)$-tensor, the curvature tensor is given by

$$R(X,Y,Z,W) = g(R_{X,Y}Z, W).$$

Let $\{e_1, \dots, e_n\}$ be an orthonormal frame. The Ricci tensor is defined as

$$\mathrm{Ric}(X,Y) = \sum_{i=1}^{n} R(X, e_i, e_i, Y),$$

and the scalar curvature is

$$\mathrm{scal} = \sum_{i=1}^{n} \mathrm{Ric}(e_i, e_i).$$

For any smooth (r,s)-tensor field T, we define

$$R_{X,Y}T = \nabla_X \nabla_Y T - \nabla_Y \nabla_X T - \nabla_{[X,Y]} T,$$

and we have the useful identity

$$\begin{aligned}
&[R_{X,Y}T](\omega_1, \dots, \omega_r, X_1, \dots, X_s) \\
&\quad = \sum_{i=1}^{r} T(\omega_1, \dots, R_{Y,X}\omega_i, \dots, \omega_r, X_1, \dots, X_s) \\
&\quad + \sum_{j=1}^{s} T(\omega_1, \dots, \omega_r, X_1, \dots, R_{Y,X}X_j, \dots, X_s).
\end{aligned}$$

We call this identity the Ricci identity. We will need this identity frequently, for instance for computing the variational formulas in the appendix. The metric induces a natural pointwise scalar product on (r, s)-tensors by

$$\langle T, S\rangle = \sum_{\substack{i_1,\ldots,i_r,\\ j_1,\ldots,j_s=1}}^{n} T(e_{i_1}^*,\ldots,e_{i_r}^*,e_{j_1},\ldots,e_{j_s})S(e_{i_1}^*,\ldots,e_{i_r}^*,e_{j_1},\ldots,e_{j_s}),$$

where $\{e_1,\ldots,e_n\}$ is an orthonormal basis at the given point and $\{e_1^*,\ldots,e_n^*\}$ is its dual basis. The global L^2-scalar product is

$$(T,S)_{L^2} = \int_M \langle T, S\rangle \, dV.$$

These induce a pointwise norm and an L^p-norm by

$$|T| := \langle T, T\rangle^{1/2},$$
$$\|T\|_{L^p} := \left(\int_M |T|^p \, dV\right)^{1/p}.$$

Furthermore, we define the C^k-norms and the Sobolev norms by

$$\|T\|_{C^k} = \sum_{i=0}^{k} \sup_{p\in M} |\nabla^i T|,$$
$$\|T\|_{W^{k,p}} = \left(\sum_{i=0}^{k} \|\nabla^i T\|_{L^p}^2\right)^{1/2}.$$

We abbreviate

$$\|T\|_{H^k} = \|T\|_{W^{k,2}},$$

and for $k \in \mathbb{N}$, we define the H^{-k}-norm as the dual norm of the H^k-norm, i.e.

$$\|T\|_{H^{-k}} = \sup_{S\neq 0} \frac{(T,S)_{L^2}}{\|S\|_{H^k}}.$$

For $\alpha \in (0,1)$ and $k \in \mathbb{N}_0$, we define the Hölder norm by

$$\|T\|_{C^{k,\alpha}} = \|T\|_{C^k} + \sup_{p\neq q} \frac{||\nabla^k T|_q - |\nabla^k T|_p|}{d(p,q)^\alpha}.$$

Given a metric, we can naturally identify vector fields and 1-forms by the map

$$P\colon \mathfrak{X}(M) \to \Omega^1(M),$$
$$X \mapsto (Y \mapsto \langle X, Y\rangle).$$

This map is called the musical isomorphism. We denote $P(X) = X^\flat$ and $P^{-1}(\omega) = \omega^\sharp$ where $X \in \mathfrak{X}(M)$, $\omega \in \Omega^1(M)$. For any $f \in C^\infty(M)$, we define the gradient as

$$\mathrm{grad} f = (\nabla f)^\sharp.$$

The Lie derivative of a smooth tensor field T along a vector field X is given by

$$\mathcal{L}_X T = \frac{d}{dt}\bigg|_{t=0} \varphi_t^* T,$$

where φ_t is the 1-parameter group of diffeomorphisms generated by X. We have the formulas

$$\begin{aligned}\mathcal{L}_X f &= X(f),\\ \mathcal{L}_X Y &= [X,Y],\\ \mathcal{L}_X \omega(Y) &= X(\omega(Y)) - \omega([X,Y]),\end{aligned}$$

where $f \in C^\infty(M)$, $X, Y \in \mathfrak{X}(M)$ and $\omega \in \Omega^1(M)$. For any (r,s)-tensor field, the above formulas extend by the Leibniz rule, i.e.

$$\begin{aligned}\mathcal{L}_X T(\omega_1,\ldots,\omega_r,X_1,\ldots,X_s) =& X(T(\omega_1,\ldots,\omega_r,X_1,\ldots,X_s))\\ &- \sum_{i=1}^r T(\omega_1,\ldots,\mathcal{L}_X\omega_i,\ldots,\omega_r,X_1,\ldots,X_s)\\ &- \sum_{i=1}^s T(\omega_1,\ldots,\omega_r,X_1,\ldots,\mathcal{L}_X X_i,\ldots,X_s).\end{aligned}$$

It is furthermore easy to see that for any metric g,

$$(\mathcal{L}_X g)(Y,Z) = g(\nabla_Y X, Z) + g(Y, \nabla_Z X),$$

where ∇ is the covariant derivative with respect to g.

By S^pM, we denote the bundle of $(0,p)$-tensors which are symmetric in all entries. We equip S^pM with the pointwise scalar product and the L^2-scalar product from above. The divergence is the map $\delta\colon \Gamma(S^pM) \to \Gamma(S^{p-1}M)$, defined by

$$\delta T(X_1,\ldots,X_{p-1}) = -\sum_{i=1}^n \nabla_{e_i} T(e_i, X_1,\ldots,X_{p-1}).$$

The adjoint map $\delta^*\colon \Gamma(S^{p-1}M) \to \Gamma(S^pM)$ with respect to the L^2-scalar product is given by

$$\delta^* T(X_1,\ldots,X_p) = \frac{1}{p}\sum_{i=0}^{p-1} \nabla_{X_{1+i}} T(X_{2+i},\ldots,X_{p+i}),$$

where the sums $1+i,\ldots,p+i$ are taken modulo p. The Laplace-Beltrami operator acting on functions (which we often just call Laplacian) is defined with the sign convention such that

$$\Delta f = -\mathrm{tr}\nabla^2 f = \nabla^*\nabla f.$$

For $T, S \in \Gamma(S^2M)$, we define their composition as

$$T \circ S(X,Y) = \sum_{i=1}^n S(X,e_i)\cdot T(e_i,Y).$$

We define an endomorphism $\mathring{R}\colon \Gamma(S^2M) \to \Gamma(S^2M)$ by

$$\mathring{R}T(X,Y) = \sum_{i=1}^{n} T(R_{e_i,X}Y, e_i).$$

Note that $\mathring{R}$ is self-adjoint with respect to the pointwise scalar product and the L^2-scalar product. For $T \in \Gamma(S^2M)$, we define the Lichnerowicz Laplacian by

$$\Delta_L T = \nabla^*\nabla T + \mathrm{Ric} \circ T + T \circ \mathrm{Ric} - 2\mathring{R}T.$$

The Lichnerowicz Laplacian is self-adjoint with respect to the L^2-scalar product.

If we wish to emphasize the dependence of the above objects on the metric, we add a g in the notation, e.g. we write R_g instead of R. For first and second variations of these objects in the direction of h, we use the notation of [Bes08], e.g. we write $R_g'(h)$, $R_g''(h)$ for the first two variations of the curvature tensor and similarly for other quantities.

Chapter 2

The Einstein-Hilbert Functional

This chapter summarizes well-known facts about the Einstein-Hilbert functional and its variational theory. The facts explained here can also be found in [Bes08; Sch89].

2.1 The Definition

Definition 2.1.1 (Einstein-Hilbert functional). Let M be a manifold and let $\mathcal{M}$ be the set of all smooth Riemannian metrics on M. The map

$$\begin{aligned} S\colon \mathcal{M} &\to \mathbb{R}, \\ g &\mapsto \int_M \mathrm{scal}_g \, dV_g \end{aligned}$$

is called Einstein-Hilbert functional. Sometimes, it is also called total scalar curvature.

As an open subset of the infinite-dimensional vector space $\Gamma(S^2M)$, $\mathcal{M}$ is an infinite-dimensional manifold. By smoothness, it cannot be modelled as a Banach manifold but as an inverse limit Hilbert manifold (ILH-manifold). In the following, we do not need details about IHL-theory so we refer the reader to [Omo68].

Remark 2.1.2. The Einstein-Hilbert functional is a Riemannian functional, i.e. for any diffeomorphism $\varphi : M \to M$, we have

$$S(\varphi^* g) = S(g),$$

where $\varphi^* g$ is the pullback metric defined by $\varphi^* g(X, Y) := g(d\varphi(X), d\varphi(Y))$.

Remark 2.1.3. In dimension 2, the Gauss-Bonnet theorem yields

$$S(g) = \int_M \mathrm{scal}_g \, dV_g = 2 \int_M K_g \, dV_g = 4\pi\chi(M),$$

where K_g is the Gaussian curvature with respect to g and $\chi(M)$ is the Euler characteristic of M. Thus, the functional is constant on $\mathcal{M}$.

2.2 First Variation

Before we compute the first variation of this functional, we remark that by compactness of M, the tangent space of $\mathcal{M}$ at any metric g is given by

$$T_g\mathcal{M} = \Gamma(S^2M).$$

Proposition 2.2.1 (First variation of the Einstein-Hilbert functional). *Let (M,g) be a Riemannian manifold. Then the first variation of the total scalar curvature in the direction of $h \in \Gamma(S^2M)$ is given by*

$$S'_g(h) = \int_M \left\langle \frac{\mathrm{scal}_g}{2} g - \mathrm{Ric}_g, h \right\rangle_g \, dV_g.$$

Proof. By the Lemmas A.1 and A.2, we have

$$\begin{aligned} \mathrm{scal}'_g(h) &= \Delta_g(\mathrm{tr}_g h) + \delta_g(\delta_g h) - \langle \mathrm{Ric}_g, h\rangle_g, \\ dV'_g(h) &= \frac{1}{2}\mathrm{tr}_g h \, dV_g. \end{aligned}$$

Therefore, by Stokes' theorem,

$$\begin{aligned} S'_g(h) &= \int_M \mathrm{scal}'_g(h) \ \ dV_g + \int_M \mathrm{scal}_g \, dV'_g(h) \\ &= \int_M [\Delta_g(\mathrm{tr}_g h) + \delta_g(\delta_g h) - \langle \mathrm{Ric}_g, h\rangle_g] \, dV_g + \frac{1}{2}\int_M \mathrm{scal}_g \cdot \mathrm{tr}_g h \, dV_g \\ &= -\int_M \langle \mathrm{Ric}_g, h\rangle_g \, dV_g + \frac{1}{2}\int_M \mathrm{scal}_g \langle g, h\rangle_g \, dV_g \\ &= \int_M \left\langle \frac{1}{2}\mathrm{scal}_g \cdot g - \mathrm{Ric}_g, h \right\rangle_g \, dV_g. \end{aligned}$$

□

Definition 2.2.2 (Einstein tensor). For a given Riemannian metric g, we define the Einstein tensor G as

$$G = \mathrm{Ric} - \frac{1}{2}\mathrm{scal} \cdot g.$$

Proposition 2.2.1 asserts that $-G$ is the L^2-gradient of the Einstein-Hilbert functional.

Corollary 2.2.3. *The critical metrics of the Einstein-Hilbert functional are the Ricci-flat metrics, i.e. the metrics satisfying* $\mathrm{Ric}_g = 0$.

Proof. By Proposition 2.2.1, the critical points are determined by the equation

$$-G_g = \frac{1}{2}\mathrm{scal}_g \cdot g - \mathrm{Ric}_g = 0.$$

By contracting, we obtain

$$\left(\frac{n}{2} - 1\right) \mathrm{scal}_g = 0$$

and since $n \geq 3$, the scalar curvature vanishes. Therefore,

$$\mathrm{Ric}_g = \mathrm{Ric}_g - \frac{1}{2}\mathrm{scal}_g \cdot g = 0.$$

Conversely, any Ricci-flat metric has vanishing Einstein tensor. □

Given some $c > 0$, we denote

$$\mathcal{M} \supset \mathcal{M}_c = \{g \in \mathcal{M} | \mathrm{vol}(M,g) = c\}.$$

This is a submanifold of $\mathcal{M}$ of codimension 1. By the variation of the volume element and by compactness, its tangent space at some metric is given by

$$T_g\mathcal{M}_c = \left\{ h \in \Gamma(S^2M) \,\middle|\, \int_M \mathrm{tr}_g h \, dV_g = 0 \right\} =: \Gamma_g(S^2M).$$

Corollary 2.2.4. *Let $g \in \mathcal{M}$ be a metric of volume c. Then g is a critical point of $S|_{\mathcal{M}_c}$ if and only if $\mathrm{Ric}_g = \mu \cdot g$ for some $\mu \in \mathbb{R}$.*

Proof. A metric g is a critical point of $S|_{\mathcal{M}_c}$ if and only if the L^2-gradient of S at g is orthogonal to $T_g\mathcal{M}_c$. This means that $G_g = \lambda \cdot g$ for some $\lambda \in \mathbb{R}$. By contraction,

$$\left(1 - \frac{n}{2}\right) \mathrm{scal}_g = \lambda \cdot n$$

and since $n \geq 3$, the scalar curvature is constant. This immediately yields

$$\mathrm{Ric}_g = G_g + \frac{1}{2}\mathrm{scal}_g \cdot g = \mu \cdot g$$

for some $\mu \in \mathbb{R}$. Conversely, if $\mathrm{Ric}_g = \mu \cdot g$, then the Einstein tensor equals $G_g = \lambda \cdot g$, where $\lambda = (1 - \frac{n}{2}) \cdot \mu$. □

Definition 2.2.5 (Einstein manifolds)**.** A Riemannian manifold (M, g) is said to be Einstein if $\mathrm{Ric}_g = \mu \cdot g$ for some $\mu \in \mathbb{R}$. We call μ the Einstein constant of g. If $\mu > 0$, we call an Einstein manifold positive, if $\mu < 0$, we call it negative. If $\mu = 0$, we call it Ricci-flat.

Remark 2.2.6. Einstein metrics also appear as the critical points of the map

$$\mathcal{M} \ni g \mapsto \mathrm{vol}(M,g)^{\frac{2}{n}-1} \int_M \mathrm{scal}_g \, dV_g$$

which is the volume-normalized variant of the Einstein-Hilbert functional. Einstein metrics with fixed constant μ are the critical points of

$$\mathcal{M} \ni g \mapsto \int_M (\mathrm{scal}_g + (2-n)\mu) \, dV_g.$$

2.3 Second Variation

To see if the Einstein-Hilbert functional has extremality properties at Einstein metrics, we now compute its second variation.

Proposition 2.3.1 (Second variation of the Einstein-Hilbert functional)**.** *Let (M, g) be an Einstein manifold with constant μ and volume c. Then the second variation of $S|_{\mathcal{M}_c}$ at g in the direction of $h \in T_g(\mathcal{M}_c)$ is given by*

$$\begin{aligned} S''_g(h) = \int_M \langle h, -\frac{1}{2}\nabla^*\nabla h + \delta^*\delta h + \delta(\delta h) g \\ + \frac{1}{2}(\Delta_g \mathrm{tr}_g h) g - \frac{\mu}{2}(\mathrm{tr}_g h) g + \mathring{R}_g h \rangle_g \; dV_g. \end{aligned} \tag{2.1}$$

Proof. Let g_t be a variation of g in $\mathcal{M}_c$ and let $h = \frac{d}{dt}|_{t=0} g_t$ and $k = \frac{d^2}{dt^2}|_{t=0} g_t$. Then by the variational formulas in Lemma A.1,

$$\begin{aligned}
\frac{d^2}{dt^2}\bigg|_{t=0} S[g_t] &= -\frac{d}{dt}\bigg|_{t=0} \int_M \langle G_{g_t}, g_t' \rangle_{g_t}\, dV_{g_t} \\
&= -\int_M \langle G_g'(h), h\rangle_g\, dV_g - \int_M \langle G_g, k\rangle_g\, dV_g \\
&\quad + 2\int_M \langle G_g, h \circ h\rangle_g\, dV_g - \frac{1}{2}\int_M \langle G_g, h\rangle_g \mathrm{tr}_g h\, dV_g.
\end{aligned}$$

Since (M, g) is Einstein, $G_g = (\frac{1}{n} - \frac{1}{2})\mathrm{scal}_g \cdot g$ and

$$\begin{aligned}
2\int_M \langle G_g, h \circ h\rangle_g\, dV_g &= 2\left(\frac{1}{n} - \frac{1}{2}\right)\mathrm{scal}_g \int_M |h|_g^2\, dV_g, \\
-\frac{1}{2}\int_M \langle G_g, h\rangle_g \mathrm{tr}_g h\, dV_g &= -\frac{1}{2}\left(\frac{1}{n} - \frac{1}{2}\right)\mathrm{scal}_g \int_M (\mathrm{tr}_g h)^2\, dV_g.
\end{aligned}$$

Since g_t is a curve in $\mathcal{M}_c$, we have

$$\begin{aligned}
0 = \frac{d^2}{dt^2}\bigg|_{t=0} \mathrm{vol}(M, g_t) &= \int_M \frac{d^2}{dt^2}\bigg|_{t=0} dV_{g_t} \\
&= \frac{1}{2}\int_M \frac{d}{dt}\bigg|_{t=0} (\mathrm{tr}_{g_t}(g_t')\, dV_{g_t}) \\
&= \frac{1}{2}\int_M [\mathrm{tr}_g k + (1/2)(\mathrm{tr}_g h)^2 - |h|_g^2]\, dV_g,
\end{aligned}$$

which implies

$$\begin{aligned}
-\int_M \langle G_g, k\rangle_g\, dV_g &= -\left(\frac{1}{n} - \frac{1}{2}\right)\mathrm{scal}_g \int_M \mathrm{tr}_g k\, dV_g \\
&= -\left(\frac{1}{n} - \frac{1}{2}\right)\mathrm{scal}_g \int_M [|h|_g^2 - \frac{1}{2}(\mathrm{tr}_g h)^2]\, dV_g.
\end{aligned}$$

By the variational formulas of the Ricci tensor and the scalar curvature (see Lemma A.2),

$$\begin{aligned}
G'(h) =& \mathrm{Ric}'(h) - \frac{1}{2}\mathrm{scal}'(h) \cdot g - \frac{1}{2}\mathrm{scal} \cdot h \\
=& \frac{1}{2}\Delta_L h - \delta^*(\delta h) - \frac{1}{2}\nabla^2 \mathrm{tr}_g h \\
& - \frac{1}{2}(\Delta_g \mathrm{tr}_g h + \delta(\delta h) - \langle \mathrm{Ric}, h\rangle_g)g - \frac{1}{2}\mathrm{scal}_g \cdot h \\
=& \frac{1}{2}\nabla^*\nabla h - \mathring{R}h - \delta^*(\delta h) - \frac{1}{2}\delta(\delta h)g - \frac{1}{2}\nabla^2 \mathrm{tr}_g h \\
& - \frac{1}{2}\Delta_g \mathrm{tr}_g h \cdot g + \frac{\mu}{2}\mathrm{tr}_g h \cdot g + (\frac{1}{n} - \frac{1}{2})\mathrm{scal}_g \cdot h,
\end{aligned}$$

which yields, after integration by parts,

$$\begin{aligned}
-\int_M \langle G'(h), h\rangle\, dV_g = \int_M \langle &-\frac{1}{2}\nabla^*\nabla h + \delta^*\delta h + \delta(\delta h)g + \frac{1}{2}\Delta_g(\mathrm{tr}_g h)g \\
&+ \mathring{R}h - \frac{\mu}{2}(\mathrm{tr}_g h)g + (\frac{1}{2} - \frac{1}{n})\mathrm{scal}_g \cdot h, h\rangle_g\, dV_g.
\end{aligned}$$

By summing up, we obtain the desired formula. □

Remark 2.3.2. The second variational formula in [Bes08, Proposition 4.55] is incorrect. There, the factor $\frac{1}{2}$ is missing in front of the $\mu(\mathrm{tr}_g h)g$-term.

2.4 A Decomposition of the Space of Symmetric Tensors

To get a better understanding of the complicated looking operator appearing in (2.1), we discuss a decomposition of the space of symmetric tensors and we consider the operator restricted to the components of the decomposition.

Lemma 2.4.1 ([Koi79b]). *For any compact Riemannian manifold (M, g), we have the following L^2-orthogonal decomposition*

$$\Gamma(S^2M) = [C^\infty(M) \cdot g + \delta_g^*(\Omega^1(M))] \oplus \mathrm{tr}_g^{-1}(0) \cap \delta_g^{-1}(0).$$

If (M, g) is an Einstein manifold but not the standard sphere, this decomposition can be refined to

$$\Gamma(S^2M) = C^\infty(M) \cdot g \oplus \delta_g^*(\Omega^1(M)) \oplus \mathrm{tr}_g^{-1}(0) \cap \delta_g^{-1}(0).$$

All these factors are infinite dimensional.

Here, $\mathrm{tr}_g^{-1}(0)$ (resp. $\delta_g^{-1}(0)$) denotes the space of tensor fields, whose trace (resp. divergence) vanishes at each point in M.

These subspaces can be interpreted geometrically as follows. Let

$$[g] = \{f \cdot g | f \in C^\infty(M), f > 0\}$$

be the conformal class of g. Then the tangent space of this submanifold of $\mathcal{M}$ at g is exactly the first factor of the decomposition above, i.e. elements of $C^\infty(M) \cdot g$ are conformal deformations of g.

Let Diff(M) be the group of diffeomorphisms on M. It has a natural right action on $\mathcal{M}$ given by $(g, \varphi) \mapsto \varphi^* g$. The action of the diffeomorphism group on g gives a submanifold $g \cdot \mathrm{Diff}(M)$ whose tangent space at g is given by the space of all Lie derivatives of the metric g. We calculate

$$\begin{aligned}
\mathcal{L}_X g(Y, Z) &= g(\nabla_Y X, Z) + g(Y, \nabla_Z X) \\
&= Y(g(X, Z)) - g(X, \nabla_Y Z) + Z(g(Y, X)) - g(\nabla_Z Y, X) \\
&= Y(X^\flat(Z)) - X^\flat(\nabla_Y Z) + Z(X^\flat(Y)) - X^\flat(\nabla_Z Y) \\
&= (\nabla_Y X^\flat)(Z) + (\nabla_Z X^\flat)(Y) \\
&= 2(\delta_g^* X^\flat)(Y, Z),
\end{aligned}$$

which shows that $T_g(g \cdot \mathrm{Diff}(M))$ is exactly $\delta_g^*(\Omega^1(M))$. This gives a description of the second factor.

Elements in $\mathrm{tr}_g^{-1}(0) \cap \delta_g^{-1}(0)$ are often called transverse traceless tensors. From now on, we abbreviate

$$TT_g = \mathrm{tr}_g^{-1}(0) \cap \delta_g^{-1}(0),$$

and we speak of TT-tensors. Deformations of the Einstein metric g in TT-directions preserve the volume element and the scalar curvature in first order.

Therefore, the third factor is often referred to as the space of non trivial volume preserving and scalar curvature preserving deformations.
From the decomposition above, we obtain

$$T_g(\mathcal{M}_c) = \Gamma_g(S^2M) = [C_g^\infty(M) \cdot g + \delta_g^*(\Omega^1(M))] \oplus TT_g \tag{2.2}$$

for Einstein metrics in general and, if g is not the standard metric on the sphere,

$$T_g(\mathcal{M}_c) = \Gamma_g(S^2M) = C_g^\infty(M) \cdot g \oplus \delta_g^*(\Omega^1(M)) \oplus TT_g. \tag{2.3}$$

Here, $C_g^\infty(M) = \{f \in C^\infty(M) | \int_M f \, dV_g = 0\}$.

Remark 2.4.2. The standard sphere is the only Einstein metric which has conformal Killing vector fields. We have

$$\frac{1}{n-1} f \cdot g_{st} = -\delta^*(\nabla f) = -2\mathcal{L}_{\mathrm{grad} f} g_{st}$$

for any $f \in C^\infty(S^n)$ with $\Delta f = n \cdot f$ where n is the smallest nonzero eigenvalue of the Laplacian. More precisely,

$$C^\infty(S^n) \cdot g_{st} \cap \delta_{g_{st}}^*(\Omega^1(S^n)) = \{f \cdot g_{st} \in C^\infty(S^n) \cdot g_{st} | \Delta f = n \cdot f\},$$

see e.g. [Oba62].

We now investigate the second variational formula of the Einstein-Hilbert functional restricted to the three components of (2.3). Let $h = f \cdot g$ for some $f \in C_g^\infty(M)$. Then (2.1) yields

$$S_g''(h) = \frac{n-2}{2} \int_M \langle f, (n-1)\Delta_g f - n\mu f \rangle \, dV_g. \tag{2.4}$$

At this point, we mention the following

Theorem 2.4.3 ([Oba62]). *Let (M, g) a compact Riemannian manifold and let λ be the smallest nonzero eigenvalue of the Laplace operator acting on $C^\infty(M)$. Assume there exists $\mu > 0$ such that* $\mathrm{Ric}(X, X) \geq \mu |X|^2$ *for any vector field X. Then λ satisfies the estimate*

$$\lambda \geq \frac{n}{n-1}\mu,$$

and equality holds if and only if (M, g) is isometric to the standard sphere.

Later on, we often refer to this theorem as Obata's eigenvalue estimate.

We conclude that $S_g''|_{C_g^\infty(M) \cdot g} \geq 0$ and $S_g''|_{C_g^\infty(M) \cdot g} > 0$ if (M, g) is not the standard sphere. Therefore, an Einstein metric is always a local minimum of the total scalar curvature restricted to metrics of the same volume in its conformal class. For the standard sphere, this is well known and for other Einstein metrics it is immediate from the strict inequality.

The second variation is easy to investigate when restricted to the second component of the splitting. Since S is a Riemannian functional, it is constant on any orbit $g \cdot \mathrm{Diff}(M)$. Because $\delta_g^*(\Omega^1(M)) = T_g(g \cdot \mathrm{Diff}(M))$, we therefore have

$$S_g''(h) = 0 \tag{2.5}$$

for each $h \in \delta_g^*(\Omega^1(M))$.

The third component appears to be the most interesting one. For $h \in TT_g$, formula (2.1) yields

$$S_g''(h) = -\frac{1}{2}\int_M \langle h, \nabla^*\nabla h - 2\mathring{R}h\rangle \, dV_g. \tag{2.6}$$

Definition 2.4.4 (Einstein Operator)**.** We call the differential operator

$$\Delta_E = \nabla^*\nabla - 2\mathring{R} \colon \Gamma(S^2M) \to \Gamma(S^2M)$$

the Einstein operator.

The Einstein operator is a self-adjoint elliptic operator. By compactness of M, $(\Delta_E + c)^{-1}$ is a compact operator on $L^2(S^2M)$ for any $c \in \mathbb{R}$ in the resolvent set of Δ_E. Therefore, by spectral theory, Δ_E has a discrete set of eigenvalues $\{\lambda_n\}$, $n \in \mathbb{N}$, forming a sequence $\lambda_1 < \lambda_2 < \ldots$, and $\lambda_n \to \infty$ as $n \to \infty$. Any eigenvalue has finite multiplicity.

The Einstein operator is closely related to the Lichnerowicz Laplacian Δ_L, which is another self-adjoint elliptic operator acting on $\Gamma(S^2M)$. In fact, on Einstein manifolds, we have the relation

$$\Delta_L = \Delta_E + 2\mu \cdot \mathrm{id}, \tag{2.7}$$

where μ is the Einstein constant of g. In addition, the Lichnerowicz Laplacian satisfies some useful properties.

Lemma 2.4.5. *Let (M, g) be a Riemannian manifold and Δ_L its Lichnerowicz Laplacian. Then*

$$\Delta_L(f \cdot g) = (\Delta f) \cdot g, \tag{2.8}$$

$$\mathrm{tr}(\Delta_L h) = \Delta(\mathrm{tr} h) \tag{2.9}$$

for all $f \in C^\infty(M)$, $h \in \Gamma(S^2M)$. Moreover, if Ric *is parallel,*

$$\Delta_L(\delta^*\omega) = \delta^*(\Delta_H \omega), \tag{2.10}$$

$$\delta(\Delta_L h) = \Delta_H(\delta h), \tag{2.11}$$

$$\Delta_L(\nabla^2 f) = \nabla^2(\Delta f) \tag{2.12}$$

for all $f \in C^\infty(M)$, $\omega \in \Omega^1(M)$, $h \in \Gamma(S^2M)$. Here, $\Delta_H = \nabla^*\nabla + \mathrm{Ric}$ is the Hodge Laplacian on 1-forms.

Proof. Formula (2.8) follows from an easy calculation. For a proof of (2.9),(2.10) and (2.12), see e.g. [Lic61, pp. 28-29]. Formula (2.11) is a consequence of (2.10). □

Lemma 2.4.6. *If g is Einstein, the Einstein operator maps TT_g to itself.*

Proof. This follows from (2.7) and Lemma 2.4.5. □

Remark 2.4.7. From (2.4), (2.5), (2.6) and Lemma 2.4.6, we conclude the following: If (M, g) is an Einstein manifold but not the standard sphere, the decomposition

$$C_g^\infty(M) \cdot g \oplus \delta_g^*(\Omega^1(M)) \oplus TT_g$$

is orthogonal with respect to the bilinear form induced by S''. In contrast, the decomposition is not L^2-orthogonal since $C_g^\infty(M)$ and $\delta_g^*(\Omega^1(M))$ are not L^2-orthogonal.

2.5 Stability of Einstein Metrics and Infinitesimal Einstein Deformations

Comparing (2.4) and (2.6), we see that an Einstein metric is neither a local minimum nor maximum of S. But it is a local minimum in its conformal class and $S''|_{TT}$ has finite coindex. This motivates the following

Definition 2.5.1 (Stability of Einstein manifolds). Let (M,g) be an Einstein manifold. We say that (M,g) is stable if its Einstein operator restricted to TT-tensors is nonnegative. If $\Delta_E|_{TT}$ is positive, we call (M,g) strictly stable. If $\Delta_E|_{TT}$ contains negative eigenvalues, we call (M,g) unstable. Furthermore, we call $\ker(\Delta_E|_{TT})$ the space of infinitesmal Einstein deformations.

Due to compactness, $\ker(\Delta_E|_{TT})$ is always finite-dimensional. In the following, we will justify the notion of infinitesimal Einstein deformations. We define an equivalence relation on $\mathcal{M}$ as follows: We call g_1 and g_2 equivalent if there exist $c>0$ and $\varphi\in\mathrm{Diff}(M)$ such that $g_2=c\cdot\varphi^*g_1$. Observe that all metrics in one equivalence class essentially contain the same geometry, since they are isometric up to rescaling. The quotiont

$$\mathcal{M}/\sim=\mathcal{M}_1/\sim$$

is called the space of all Riemannian structures. The quotient of the set of all Einstein metrics under this relation is called the moduli space of Einstein structures. A local description of the set of Riemannian structures is given by the slice theorem. For us, the following parts of the theorem are important (see also [Bes08, p.345] for a more detailed formulation).

Theorem 2.5.2 (Ebin). *Let g_0 be a unit-volume Riemannian metric on a compact manifold M. Then there exists a submanifold $\mathcal{S}_{g_0}\subset\mathcal{M}_1$ with tangent space*

$$T_{g_0}\mathcal{S}_{g_0}=T_g\mathcal{M}_1\cap\delta_{g_0}^{-1}(0)$$

and a neighbourhood $\mathcal{U}\subset\mathcal{M}_1$ of g_0 such that for any $g\in\mathcal{U}$, there exist $\bar{g}\in\mathcal{S}_{g_0}$ and $\varphi\in\mathrm{Diff}(M)$ such that $g=\varphi^*\bar{g}$. We call $\mathcal{S}_{g_0}$ a slice of the action of $\mathrm{Diff}(M)$.

The theorem basicly says that all geometries close to g_0 are contained in the slice $\mathcal{S}_{g_0}$. Due to rescaling, the analogous assertion of course holds for manifolds with arbitrary volume.

Now, let g_t be a curve of Einstein metrics of volume c through $g=g_0$ lying in the slice $\mathcal{S}_g$. Then, since all g_t are critical points of the Einstein-Hilbert functional restricted to $\mathcal{M}_c$, the function $t\mapsto S(g_t)=\mathrm{vol}(M,g_t)\cdot\mathrm{scal}_{g_t}$ is constant. We immediately obtain that the scalar curvature (and hence the Einstein constant) is constant in t and that $h=\frac{d}{dt}|_{t=0}g_t$ satisfies the system

$$\delta_g h=0,\qquad \int_M \mathrm{tr}_g h\, dV_g=0,\qquad \left.\frac{d}{dt}\right|_{t=0}\left(\mathrm{Ric}_{g_t}-\frac{S(g_t)}{c\cdot n}\cdot g_t\right)=0. \tag{2.13}$$

By a result of Berger and Ebin (see [Bes08, Theorem 12.30]), this is equivalent to the system

$$\delta_g h=0,\qquad \mathrm{tr}_g h=0,\qquad \Delta_E h=0. \tag{2.14}$$

In other words, h is an infinitesimal Einstein deformation. We conclude that if $\ker(\Delta_E|_{TT})$ is trivial, the Einstein metric g is isolated in the moduli space of Einstein structures, i.e. there are no other Einstein metrics close to g except those of the form $c \cdot \varphi^* g$. On the other hand, it is in general not true that for any $h \in \ker(\Delta_E|_{TT})$, there exists a curve of Einstein metrics tangent to h. In fact, g can be isolated in the moduli space although $\ker(\Delta_E|_{TT})$ is nontrivial. Such examples (e.g. the product metric on $\mathbb{C}P^{2n} \times S^2$) are discussed in [Koi82].

Definition 2.5.3. An infinitesimal Einstein deformation h is said to be integrable if there exists a curve g_t of Einstein metrics such that $\frac{d}{dt}|_{t=0} g_t = h$.

Remark 2.5.4. Stability properties of compact Riemannian Einstein metrics also play a role in mathematical general relativity. In [AM11], L. Andersson and V. Moncrief consider the Lorentzian cone over a compact negative Einstein manifold. They prove a global existence theorem for solutions of Einstein's equations close to the cone under the assumption that the compact Einstein metric is stable.

Remark 2.5.5. The eigenvalues of the Einstein operator (resp. the Lichnerowicz Laplacian) acting on TT-tensors are also important for the stability of higher-dimensional black holes and event horizons in physics, see [GH02; GHP03]. There, a stability conditon (which we may call physical stability) on Einstein manifolds with constant $\mu > 0$ is given by

$$\lambda \geq -\frac{\mu}{n-1}\left(4 - \frac{1}{4}(n-1)^2\right)$$

for the smallest eigenvalue of the Lichnerowicz Laplacian acting on TT-tensors.

2.6 The Manifold of Metrics of constant Scalar Curvature

On a given compact manifold, there exist many metrics of constant scalar curvature. We introduce the notations

$$\begin{aligned} \mathcal{C} &= \{g \in \mathcal{M} | \mathrm{scal}_g \text{ is constant}\}, \\ \mathcal{C}_c &= \mathcal{M}_c \cap \mathcal{C}. \end{aligned}$$

Let $\Psi \colon \mathcal{M} \to C^\infty(M)$ be defined by $\Psi(g) = \Delta_g \mathrm{scal}_g$. Since M is compact, $\mathcal{C} = \Psi^{-1}(0)$. Let $g \in \mathcal{C}$. By the first variation of the scalar curvature, the differential of Ψ at g is equal to

$$\alpha_g(h) = d\Psi_g(h) = \Delta_g(\mathrm{scal}'_g(h)) = \Delta_g(\Delta_g \mathrm{tr}_g h + \delta_g(\delta_g h) - \langle \mathrm{Ric}, h\rangle). \quad (2.15)$$

Theorem 2.6.1 ([Koi79a]). *Let $g_0 \in \mathcal{C}_1$ such that $\mathrm{scal}_{g_0}/(n-1)$ is not a positive eigenvalue of the Laplace-Beltrami operator. Then in a neighbourhood of g_0, $\mathcal{C}_1$ is an ILH-submanifold of $\mathcal{M}$ such that*

$$T_{g_0}\mathcal{C}_1 = \ker(\alpha_{g_0}) \cap \left\{ h \in \Gamma(S^2M) \,\middle|\, \int_M \mathrm{tr}_{g_0} h \, dV_{g_0} = 0 \right\}. \quad (2.16)$$

Furthermore, the map $(f, g) \mapsto f \cdot g$ from $C^\infty(M) \times \mathcal{C}_1$ to $\mathcal{M}$ is a local ILH-diffeomorphism from a neighbourhood of $(1, g_0)$ to a neighbourhood of g_0.

By rescaling, this local decomposition holds of course for constant scalar curvature metrics of arbitrary volume. Observe that this assertion holds for all Einstein metrics except the standard sphere. The local decomposition follows from the ILH inverse function theorem and the splitting

$$\Gamma(S^2M) = C^\infty(M) \cdot g \oplus T_g\mathcal{C}_c.$$

If g is an Einstein metric, then $g \cdot \mathrm{Diff}(M) \subset \mathcal{C}_c$, and therefore,

$$\delta^*(\Omega^1(M)) = T_g(g \cdot \mathrm{Diff}(M)) \subset T_g\mathcal{C}_c.$$

By (2.15) and (2.16), it is easy to see that $TT_g \subset T_g\mathcal{C}_1$. Since also the decomposition

$$\Gamma(S^2M) = C^\infty(M) \cdot g \oplus \delta_g^*(\Omega^1(M)) \oplus TT_g$$

holds, we have

$$T_g\mathcal{C}_c = \delta^*(\Omega^1(M)) \oplus TT_g,$$

and in addition, this decomposition is L^2-orthogonal.

Proposition 2.6.2 ([Bes08],Proposition 4.47). *Let (M, g) be a metric of constant scalar curvature and volume c such that $\mathrm{scal}_g/(n-1) \notin \mathrm{spec}_+(\Delta_g)$. Then g is a critical point of $S|_{\mathcal{C}_c}$ if and only if g is Einstein.*

Observe that stability of an Einstein manifold precisely means that the second variation of the Einstein-Hilbert functional is nonpositive on $T_g\mathcal{C}_c$. The following lemma is quite immediate but is not stated in this form in the literature.

Lemma 2.6.3. *Let (M, g_0) be an Einstein manifold of volume c. If we have $\mathrm{scal}_g \leq \mathrm{scal}_{g_0}$ for all $g \in \mathcal{C}_c$ in a small C^2-neighbourhood of g_0, then (M, g_0) is stable. Conversely, if (M, g_0) is strictly stable, there exists a C^2-neighbourhood $\mathcal{U}$ of g_0 in the space of metrics such that $\mathrm{scal}_g \leq \mathrm{scal}_{g_0}$ for all $g \in \mathcal{U} \cap \mathcal{C}_c$, and equality holds if and only if g is isometric to g_0.*

Proof. Suppose that g_0 is unstable, then there exists $h \in TT_g \subset T_g\mathcal{C}_c$ such that $S''_{g_0}(h) > 0$. By integrating, we obtain a curve $g_t \in \mathcal{C}_c$ such that we have $\mathrm{scal}_{g_t} = c^{-1}S(g_t) > c^{-1}S(g_0) = \mathrm{scal}_{g_0}$ for all $t \in (0, \epsilon)$. Conversely, suppose that (M, g_0) is strictly stable, i.e. S'' is negative on TT-tensors. Let $\mathcal{S}_{g_0}$ be a slice through g_0 and consider the set $\mathcal{S}_{g_0} \cap \mathcal{C}_c$. This is an infinite-dimensional submanifold of $\mathcal{M}$ and its tangent space through g_0 is given by TT. The map $g \mapsto \mathrm{scal}_g = c^{-1}S[g]$ is a smooth functional on $\mathcal{S}_{g_0} \cap \mathcal{C}_c$ which is continuous with respect to the C^2-topology. Since g_0 is a critical point of scal and the second variation is negative at g_0, there is a small C^2-neighbourhood $\mathcal{V} \subset \mathcal{S}_{g_0} \cap \mathcal{C}_c$ such that $\mathrm{scal}_g < \mathrm{scal}_{g_0}$ for all $g \in \mathcal{V}$, $g \neq g_0$. By the slice theorem, there exists a C^2-neighbourhood $\mathcal{U}$ in $\mathcal{M}$ such that any $g \in \mathcal{C}_c \cap \mathcal{U}$ can be written as $\varphi^*\bar{g}$ for some $\varphi \in \mathrm{Diff}(M)$ and $\bar{g} \in \mathcal{V}$. By diffeomorphism invariance, $\mathrm{scal}_g = \mathrm{scal}_{\bar{g}} \leq \mathrm{scal}_{g_0}$ and equality holds if and only if $g = \varphi^* g_0$. □

2.7 The Yamabe Invariant

For a smooth metric g on a given compact manifold, we consider

$$Y(M,[g]) = \inf_{\tilde{g}\in[g]} \mathrm{vol}(M,\tilde{g})^{(2-n)/n} \int_M \mathrm{scal}_{\tilde{g}}\, dV_{\tilde{g}},$$

where $[g]$ is the conformal class of g. We call this infimum the Yamabe constant of the conformal class of g. By the solution of the Yamabe problem (which was solved by Schoen in [Sch84]), it is well known that this infimum is always finite and that it is realized by a metric of constant scalar curvature. Metrics realizing this infimum are nessecarily of constant scalar curvature and are called Yamabe metrics. We now define the Yamabe functional

$$\begin{aligned} Y\colon \mathcal{M} &\to \mathbb{R}, \\ g &\mapsto Y(M,[g]). \end{aligned}$$

By definition, this functional is conformally invariant. It is also a diffeomorphism invariant, so $Y(\varphi^* g) = Y(g)$ for any $\varphi \in \mathrm{Diff}(M)$. The Yamabe functional is continuous with respect to the C^2-topology (see [Bes08, Proposition 4.31]). We call

$$Y(M) = \sup_{g\in\mathcal{M}} Y(M,[g])$$

the Yamabe invariant of M. It is well known (see e.g. [LP87, p. 50]) that

$$Y(M,[g]) \leq Y(S^n,[g_{st}]),$$

and equality holds if and only if $M = S^n$ and g is isometric to some metric in $[g_{st}]$. Thus, we immediately obtain the bound

$$Y(M) \leq Y(S^n).$$

In particular, the Yamabe invariant of any manifold is always a real number.

Definition 2.7.1. A Yamabe metric g is called supreme if $Y(M,[g]) = Y(M)$.

Let $\mathcal{Y}_c$ be the set of Yamabe metrics of volume c. Clearly, $\mathcal{Y}_c \subset \mathcal{C}_c$. Let g be an Einstein metric. It is well-known that any Einstein metric is the unique Yamabe metric in its conformal class (see [Sch89, Proposition 1.4]).

Theorem 2.7.2 ([BWZ04]). *Let (M, g_0) be an Einstein metric not conformally equivalent to the round sphere. Then any metric $g \in \mathcal{C}$ which is $C^{2,\alpha}$-close to g_0 is a Yamabe metric.*

In other words, $\mathcal{U} \cap \mathcal{Y}_c = \mathcal{U} \cap \mathcal{C}_c$ for a small $C^{2,\alpha}$-neighbourhood $\mathcal{U}$ of g_0. Moreover, if $\mathcal{U}$ is small enough, Theorem 2.6.1 implies that a small $C^{2,\alpha}$-neighbourhood of $(1, g_0)$ in $C^\infty(M) \times \mathcal{C}_c$ is mapped diffeomorphically to $\mathcal{U}$ by $(f,g) \mapsto f\cdot g$. Since g is Yamabe, we have

$$Y(f\cdot g) = Y(g) = c^{2/n}\cdot \mathrm{scal}_g$$

for $f\cdot g \in \mathcal{U}$. This shows that the Yamabe functional is smooth on $\mathcal{U}$, since $g \mapsto \mathrm{scal}_g$ is smooth on $\mathcal{C}_c$. Using these observations, we can deduce the following from Proposition 2.6.2:

Corollary 2.7.3. *Let (M, g_0) be an Einstein manifold. If $Y(g) \leq Y(g_0)$ for all $g \in \mathcal{M}$ in a small $C^{2,\alpha}$-neighbourhood of g_0, then (M, g_0) is stable. Conversely, if (M, g_0) is strictly stable, there exists a $C^{2,\alpha}$-neighbourhood $\mathcal{U}$ of g_0 in the space of metrics such that $Y(g) \leq Y(g_0)$ for all $g \in \mathcal{U}$, and equality holds if and only if g is isometric to g_0.*

In particular, we have

Corollary 2.7.4. *Any supreme Einstein metric (M, g) is stable.*

Chapter 3

Some stable and unstable Einstein Manifolds

In this chapter, we study the Einstein operator on particular examples. In the first section, we mention some well-known examples and classes of stable and unstable Einstein manifolds. In Section 3.2, we study the Einstein operator on Bieberbach manifolds and we compute the dimension of its kernel in terms of the holonomy. In Section 3.3, we study the Einstein operator on products of Einstein spaces.

3.1 Standard Examples

In general, it is very hard to find out if an Einstein manifold is stable or not. However for some examples, this is possible and for very few examples, it is even possible to compute the spectrum of the Einstein operator explicitly.

Example 3.1.1 (The flat torus). We consider the Torus $T^n = \mathbb{R}^n/\mathbb{Z}^n$ equipped with the flat metric. We consider the Einstein operator acting on the subbundle $\mathrm{tr}^{-1}(0) \subset \Gamma(S^2M)$. This is a trivial vector bundle over T^n and its dimension equals $n(n+1)/2-1$. Since the manifold is flat, $\Delta_E = \nabla^*\nabla = \Delta_0 \oplus \ldots \oplus \Delta_0$ where Δ_0 is the usual Laplace-Beltrami operator acting on functions. Therefore, the spectrum of Δ_E concides with the spectrum of the Laplace-Beltrami operator on T^n, so

$$\mathrm{spec}(\Delta_E|_{\mathrm{tr}^{-1}(0)}) = \left\{(2\pi)^2(k_1^2+\ldots+k_n^2)|\; k_i \in \mathbb{Z}\right\}.$$

In particular, since all eigenvalues are nonnegative, (T^n, g_{eukl}) is stable. The kernel of $\Delta_E|_{\mathrm{tr}^{-1}(0)}$ has dimension $n(n+1)/2-1$, since it consists precisely of the parallel sections in $\mathrm{tr}^{-1}(0)$, which are obviously TT-tensors.

Example 3.1.2 (The sphere and the real projective space). The n-dimensional unit sphere is an Einstein manifold with constant $(n-1)$. Here, we have the relation

$$\Delta_E = \Delta_L - 2(n-1),$$

where Δ_L is the Lichnerowicz Laplacian. The spectrum of Δ_L on the standard sphere was explicitly computed in [Bou99]. For Δ_L acting on traceless transverse tensors, eigentensors were constructed in the proof of [Bou99, Proposition 3.19] and the spectrum is given in [Bou99, Theorem 3.2]. We immediately obtain the spectrum of Δ_E:

$$\mathrm{spec}(\Delta_E|_{TT}) = \{k(k+n-1)|\ k \geq 2\}\,.$$

In particular, (S^n, g_{st}) is strictly stable since all eigenvalues are positive. The real projective space $(\mathbb{R}P^n, g_{st})$ is also strictly stable. The spectrum of its Einstein operator is

$$\mathrm{spec}(\Delta_E|_{TT}) = \{2k(2k+n-1)|\ k \geq 2\}\,,$$

see [Bou99, Theorem 4.2].

Example 3.1.3 (Coverings). Let $\varphi : (\tilde{M}, \tilde{g}) \to (M, g)$ be a finite Riemannian covering of Einstein manifolds. If $(\tilde{M}, \tilde{g})$ is (strictly) stable then (M, g) is (strictly) stable. This is due to the fact that any TT-eigentensor of Δ_E on M can be lifted to a TT-eigentensor of Δ_E on $\tilde{M}$ with the same eigenvalue.

Example 3.1.4 (Symmetric spaces of compact type). For most symmetric spaces of compact type, it is known if they are stable or not. A table collecting the smallest eigenvalue of the Lichnerowicz Laplacian (from which we obtain the smallest eigenvalue of the Einstein operator immediately) on such spaces is given in [CH13, p.15-17]. The only known unstable manifolds in this class are $Spin(5)$, $Sp(\mathrm{n})$, $n \geq 3$, $SO(5)/(SO(3) \times SO(2))$ and $Sp(\mathrm{n})/U(\mathrm{n})$, $n \geq 3$. For $\mathbb{H}P^2$ and $Sp(\mathrm{p{+}q})(Sp(\mathrm{p}) \times Sp(\mathrm{q}))$, it is not known whether they are stable or not. All other manifolds in this class are known to be stable. From these, the spaces $SU(\mathrm{n})$, $n \geq 3$, $SU(\mathrm{n})/O(\mathrm{n})$, $n \geq 3$, $SU(2\mathrm{n})/Sp(\mathrm{n})$, $n \geq 3$, $U(\mathrm{p{+}q})(U(\mathrm{p}) \times U(\mathrm{q}))$, $p, q \geq 2$ and E_6/F_4 have infinitesimal Einstein deformations.

Example 3.1.5 (Spin manifolds). Suppose now that our manifold (M, g) is spin. We call a nonzero spinor σ a real Killing spinor, if $\nabla_X \sigma = cX \cdot \sigma$ for some $c \in \mathbb{R}$. Any Riemannian manifold carrying a real Killing spinor is Einstein with constant $4c^2(n-1)$. If $c = 0$, σ is parallel and (M, g) is Ricci-flat. By the work in [Wan91], [DWW05], it is known that such manifolds are stable. The idea is as follows: Given a real Killing spinor, we associate to each symmetric $(0,2)$-tensor a spinor-valued 1-form by

$$\begin{aligned}\Psi\colon \Gamma(S^2M) \to& \Gamma(T^*M \otimes S),\\ h \mapsto& (X \mapsto h(X) \cdot \sigma).\end{aligned}$$

Here, S denotes the spinor bundle of (M, g) and h is considered as an endomorphism on TM. Now a straightforward calculation shows that if $h \in TT$,

$$D^2 \circ \Psi(h) + 2cD \circ \Psi(h) = \Psi \circ \Delta_E(h) + c^2 n(n-1)\Psi(h).$$

where D is the twisted Dirac operator acting on $\Gamma(T^*M \otimes S)$. Since D^2 is a nonnegative operator, (M, g) is stable if $c = 0$.

For the more general case of non-parallel real Killing spinors, stability can not be derived. We then replace the connection on $T^*M \otimes S$ by a new connection $\tilde{\nabla}$ defined by $\tilde{\nabla}_X = \nabla_X + \frac{c}{n} X\cdot$ and obtain the Bochner formula

$$\Psi \circ \Delta_E(h) = \tilde{D}^2 \circ \Psi(h) - c^2(n-1)^2 \Psi(h),$$

where $\tilde{D}$ is the Dirac operator associated to $\tilde{\nabla}$. Thus, the smallest eigenvalue of $\Delta_E|_{TT}$ is bounded from below by $-\frac{n-1}{4}\mu$ where $\mu > 0$ is the Einstein constant. As we will see, this estimate is rather bad compared to the ones discussed in the next chapter.

Example 3.1.6 (Kähler manifolds)**.** Kähler-Einstein manifolds with nonpositive Einstein constant are stable. This will be discussed in more detail in Section 4.6.

Example 3.1.7 (Product manifolds)**.** The prototypical example of an unstable Einstein manifold is the product of two positive Einstein manifolds (M^{n_1}, g_1), (N^{n_2}, g_2). Take $h = n_2 \cdot g_1 - n_1 \cdot g_2$. Then $h \in TT_{g_1+g_2}$ and $\Delta_E h = -2\mu h$ where μ is the Einstein constant.

Example 3.1.8 (Other unstable manifolds)**.** From the estimates in [GH02; GHP03; GM02; PP84a; PP84b], the following examples are also unstable:

- The three infinite families of homogeneous Einstein metrics in dimensions 5 and 7 in [Rom85; CDF84; DFVN84; PP84b]. They are S^1 -bundles over $S^2 \times S^2$, $\mathbb{C}P^2 \times S^2$ and $S^2 \times S^2 \times S^2$, respectively. These examples are special cases of the examples in [WZ86].
- A few of the inhomogeneous Einstein metrics on the products of spheres in low dimensions constructed by C. Böhm in [Böh98].

In [Böh05], C. Böhm constructed unstable Einstein metrics on the total spaces of principal torus bundles over products of Kähler-Einstein manifolds.

All the unstable Einstein metrics from the previous examples have positive scalar curvature. In contrast, no unstable Einstein metrics with nonpositive scalar curvature are known (in the compact case). This raises the following

Question ([KW75; Dai07])**.** Are all compact Einstein manifolds with nonpositive scalar curvature stable?

This is not true in the noncompact case since the Riemannian Schwarzschild metric is unstable (see [GPY82, Sec. 5]).

3.2 Bieberbach Manifolds

Bieberbach manifolds are flat connected compact manifolds. It is well known that any Bieberbach manifold is isometric to $\mathbb{R}^n/G$, where G is a suitable subgroup of the Euclidean motions $E(n) = O(n) \ltimes \mathbb{R}^n$. We call such groups Bieberbach groups. For every element $g \in E(n)$, there exist unique $A \in O(n)$ and $a \in \mathbb{R}^n$ such that $gx = Ax + a$ for all $x \in \mathbb{R}^n$, and we write $g = (A, a)$. There exist homomorphisms $r\colon E(n) \to O(n)$ and $t\colon \mathbb{R}^n \to E(n)$, defined by $r(A, a) = A$ and $t(a) = (1, a)$. Let G be a Bieberbach group. The subgroup $r(G) \subset O(n)$ is called the holonomy of G since its natural representation on $\mathbb{R}^n$ is equivalent to the holonomy representation of $\mathbb{R}^n/G$ (see e.g. [Cha86]).

We call two Bieberbach manifolds M_1 and M_2 affinely equivalent if there exists a diffeomorphism $F : M_1 \to M_2$ whose lift to the universal coverings $\pi_1\colon \mathbb{R}^n \to M_1$, $\pi_2\colon \mathbb{R}^n \to M_2$ is an affine map $\alpha\colon \mathbb{R}^n \to \mathbb{R}^n$ such that

$$\pi_2 \circ \alpha = F \circ \pi_1.$$

If M_1 and M_2 are affinely equivalent, the corresponding Bieberbach groups G_1 and G_2 are isomorphic via $\varphi : G_1 \to G_2$, $\varphi(g) = \alpha g \alpha^{-1}$. Conversely, if two Bieberbach groups G_1 and G_2 are isomorphic, there exists an affine map α such that the isomorphism is given by $g \mapsto \alpha g \alpha^{-1}$ (see [Wol11, Theorem 3.2.2]). The map α descends to a diffeomorphism $F : M_1 \to M_2$ and M_1 and M_2 are affinely equivalent via F.

Now we want to determine whether a Bieberbach manifold has infinitesimal Einstein deformations or not. Any Bieberbach manifold is stable since

$$(\Delta_E h, h)_{L^2} = (\nabla^* \nabla h, h)_{L^2} = \|\nabla h\|_{L^2}^2 \geq 0.$$

Furthermore, we see that any infinitesimal Einstein deformation is parallel.

Remark 3.2.1. The following lemma is a consequence of the holonomy principle. It also follows immediately from [Die13, Proposition 4.2].

Lemma 3.2.2. *Let (M, g) be a connected Riemannian manifold. There exists a nonzero traceless symmetric $(0,2)$-tensor field h with $\nabla h \equiv 0$ if and only if the holonomy of (M, g) is reducible.*

Proof. Let h be a symmetric and traceless parallel tensor field and consider it as an endomorphism $h : TM \to TM$. Let $p \in M$ and let $\{e_1, \ldots, e_n\}$ be a local orthonormal frame around p such that all e_i are eigenvectors of h at each point, i.e. $h(e_i) = \lambda_i e_i$. Since h is parallel,

$$\begin{aligned}
0 =& \nabla_{e_i}(h(e_j)) - h(\nabla_{e_i} e_j) \\
=& \nabla_{e_i}(\lambda_j e_j) - \sum_{k=1}^{n} h(\Gamma_{ij}^k e_k) \\
=& (\nabla_{e_i}\lambda_j) e_j + \lambda_j(\nabla_{e_i} e_j) - \sum_{k=1}^{n} \lambda_k \Gamma_{ij}^k e_k \\
=& (\nabla_{e_i}\lambda_j) e_j + \sum_{k=1}^{n} (\lambda_j - \lambda_k)\Gamma_{ij}^k e_k \\
=& (\nabla_{e_i}\lambda_j) e_j + \sum_{k=1, k\neq j}^{n} (\lambda_j - \lambda_k)\Gamma_{ij}^k e_k.
\end{aligned}$$

Since the e_i are linearly independant, $\nabla_{e_i}\lambda_j = 0$ for all $1 \leq i, j \leq n$. Thus, the eigenvalues λ_i are constant as functions on M. Let now $\lambda_1 < \ldots < \lambda_k$ be the pairwise distinct eigenvalues of h. We obtain an orthogonal splitting

$$TM = E(\lambda_1) \oplus \ldots \oplus E(\lambda_k).$$

where $E(\lambda_i)$ is the space of eigensections to the eigenvalue λ_i. Since h is trace-free, there exist at least two distinct eigenvalues, so the splitting into eigenspaces is nontrivial. Let now $\gamma : [0,1] \to M$ be a piecewise smooth curve. Let $X_0 \in T_{\gamma(0)}M$ and X_t, $t \in [0,1]$ be the parallel translated vector field along γ. Then $h(X_t)$ is the parallel translated vector field of $h(X_0)$ along γ. Therefore, the eigenspaces of h are preserved by parallel translation along curves, so the holonomy representation on T_pM leaves the eigenspaces of h invariant. Thus, it is reducible.

To prove the converse, suppose that the holonomy of (M,g) is reducible. Let $p \in M$ and $(E_1)_p, \ldots (E_k)_p$ be invariant subspaces of $Hol_p(M,g)$ such that $(E_1)_p \oplus \ldots \oplus (E_k)_p = T_pM$. Since $Hol_p(M,g) \subset O(T_pM, g_p)$, this sum is orthogonal. By parallel translation of the $(E_i)_p$, we obtain a well-defined parallel splitting $E_1 \oplus \ldots \oplus E_k = TM$. The metric splits as $g = g_1 \oplus \ldots \oplus g_k$ since these subbundles are orthogonal. Then any combination $\sum_{i=1}^k \lambda_i g_i$, $\lambda_i \in \mathbb{R}$ is a symmetric parallel tensor field and its trace vanishes for a suitable choice of the λ_i. □

Corollary 3.2.3. *A Bieberbach manifold $M = \mathbb{R}^n/G$ is strictly stable if and only if the subgroup $r(G) \subset O(n)$ acts irreducibly on $\mathbb{R}^n$.*

Proof. Recall that $r(G)$ is isomorphic to the holonomy of M. Since any infinitesimal Einstein deformation is parallel, the assertion is immediate from Lemma 3.2.2. □

We now consider the contrary case where the holonomy is reducible. We then know that the space of infinitesimal Einstein deformations is nontrivial. We want to compute its dimension. Let $TM = E_1 \oplus \ldots \oplus E_k$ be a parallel splitting of the tangent bundle into irreducible components. Then a parallel splitting of the bundle of symmetric $(0,2)$-tensors is given by

$$T^*M \odot T^*M = \bigoplus_{i,j=1}^{k} E_i^* \odot E_j^* = \bigoplus_{i=1}^{k} \odot^2 E_i^* \oplus \bigoplus_{i<j}^{k} E_i^* \odot E_j^*. \tag{3.1}$$

Here, E_i^* is the image of E_i under the musical isomorphism and $\odot$ denotes the symmetric tensor product. We now want to determine the space of parallel sections in each of these summands. First suppose that $h \in \Gamma(\odot^2 E_i)$ is parallel. Considered as an endomorphism on TM, it induces an endomorphism $h : E_i \to E_i$. By the proof of Lemma 3.2.2, its eigensections form a splitting of the bundle E_i. Since we assumed E_i to be irreducible, there can only exist one eigenvalue, which implies that $h = \lambda g_i$ where $\lambda \in \mathbb{R}$ and g_i is the metric restricted to E_i. Thus, parallel tensors in the component $\bigoplus_{i=1}^k \odot^2 E_i^*$ are of the form

$$h = \sum_{i=1}^{k} \lambda_i g_i, \qquad \lambda_i \in \mathbb{R}.$$

If we assume h to be trace-free, we have the condition

$$\sum_{i=1}^{k} \lambda_i \mathrm{dim}(E_i) = 0.$$

We have just obtained a $k-1$-dimensional space of infinitesimal Einstein deformations. Now we consider the second component of the splitting (3.1). Sections of $E_i^* \odot E_j^*$, considered as endomorphisms on TM, are sections of $\mathrm{End}(E_i \oplus E_j)$ which are of the form

$$h = \begin{pmatrix} 0 & A^* \\ A & 0 \end{pmatrix},$$

where $A \in \Gamma(\mathrm{End}(E_i, E_j))$ and A^* is its adjoint. Any such map is trace-free. Now if h is parallel, A is also parallel. Therefore, $\ker(A)$ and $\mathrm{im}(A)$ are both parallel subbundles of E_i, E_j, respectively. Since E_i, E_j are irreducible, this shows that A is an isomorphism if it is nonzero. We now want to state nessecary and sufficient conditions which ensure the existence of such a map.

Fix a point p and consider a linear map $A_p : (E_i)_p \to (E_j)_p$. It is clear that there exists at most one parallel endomorphism $A : E_i \to E_j$ which coincides with A_p at p. Assume that such an A exists and let $\gamma : [0,1] \to M$ be a closed curve starting and ending at p. Then A commutes with the parallel transport along γ and therefore, A_p commutes with the holonomy representation $\rho(Hol_p(M,g)) \subset O(T_pM, g_p)$. This is also a sufficient condition. If A_p commutes with the holonomy representation, one obtains a well-defined parallel endomorphism A by parallel translation along curves.

This condition precisely means that the restricted standard holonomy representatios $\rho(Hol_p(M,g))|_{E_i}$ and $\rho(Hol_p(M,g))|_{E_j}$ are equivalent via A_p. Recall that two representations $\rho_1 : G \to L(V)$, $\rho_2 : G \to L(W)$ are equivalent if there exists an isomorphism $\varphi : V \to W$ such that $\rho_2(g) \circ \varphi = \varphi \circ \rho_1(g)$ for all $g \in G$.

Since the representations $\rho(Hol_p(M,g))|_{E_i}$ and $\rho(Hol_p(M,g))|_{E_j}$ are finite dimensional and irreducible, the space of linear maps $L : (E_i)_p \to (E_j)_p$ commuting with these representations is 1-dimensional. This follows easily from a Lemma from representation theory (see e.g. [NS82, p. 27]).

In summary, we have shown that the dimension of the space of parallel sections in $E_i^* \odot E_j^*$ equals 1 if the holonomy representations restricted to E_i and E_j are equivalent and zero otherwise. Summing over all $E_i^* \odot E_j^*$, $i < j$ and using the fact that the representations $\rho : Hol_p(M,g) \to O(T_pM, g_p)$ and $r : G \to O(n)$ are equivalent, we obtain

Proposition 3.2.4. *Let $(M = \mathbb{R}^n/G, g)$ be a Bieberbach manifold and let ρ be the canonical representation of the subgroup $r(G)$ on $\mathbb{R}^n$. Let*

$$\rho \cong (\rho_1)^{i_1} \oplus \ldots \oplus (\rho_l)^{i_l}$$

be an irreducible decomposition of ρ. Then the dimension of infinitesimal Einstein deformations is equal to

$$\dim(\ker(\Delta_E|_{TT}) = -1 + \sum_{j=1}^{l} i_j + \sum_{j=1}^{l} \frac{i_j(i_j-1)}{2}.$$

Remark 3.2.5. We show that each of the infinitesimal Einstein deformations above is integrable. Let $M = \mathbb{R}^n/G$ be a Bieberbach manifold with the flat metric g and let h be a parallel tensor field. For small values of t, $g_t = g + th$ is also a metric. Choose local coordinates such that $g_{ij} = \delta_{ij}$. Then the local coefficients h_{ij} are constant, since h is parallel. Thus, also g_t has constant coefficients with respect to these coordinates, which implies that the Riemann curvature tensor vanishes. In particular, (M, g_t) is a curve of Einstein metrics.

Recall that two Bieberbach manifolds M_1 and M_2 are called affinely equivalent if there exists a diffeomorphism $F : M_1 \to M_2$ whose lift to the universal coverings $\pi_1 \colon \mathbb{R}^n \to M_1$, $\pi_2 \colon \mathbb{R}^n \to M_2$ is an affine map $\alpha \in GL(n) \ltimes \mathbb{R}^n$ such that $F \circ \pi_1 = \pi_2 \circ \alpha$. Since π_1, π_2 are local isometries and α is affine, the map F is parallel, i.e.

$$\nabla^{M_2}_X dF(Y) = dF(\nabla^{M_1}_{dF^{-1}(X)} Y), \qquad \forall X \in \mathfrak{X}(M_2), Y \in \mathfrak{X}(M_1).$$

The map F induces an ismorphism $F_*\colon \Gamma(S^2M_1) \to \Gamma(S^2M_2)$ which is defined as $F_*h(X,Y) = h(dF^{-1}(X), dF^{-1}(Y))$. Since F is parallel,

$$\nabla_X^{M_2} F_*h = F_*\nabla^{M_1}_{dF^{-1}(X)}h, \qquad \forall X \in \mathfrak{X}(M_2).$$

Therefore, F_* maps parallel tensor fields on M_1 isomorphically to parallel tensor fields on M_2. It follows that the dimension of infinitesimal Einstein deformations only depends on the affine equivalence class of M.

For any $n \in \mathbb{N}$ the number of affine equivalence classes of n-dimensional Bieberbach manifolds is finite (see [Bie12]). In dimension 3, a classification of all Bieberbach manifolds up to affine equivalence is known. In fact, there exist 10 Bieberbach 3-manifolds where six of them are orientable and the others are non-orientable. We describe the corresponding Bieberbach groups in the following. Moreover, we will compute the dimension of infinitesimal Einstein deformations explicitly. Let $\{e_1, e_2, e_3\}$ be the standard basis of $\mathbb{R}^3$, let $R(\varphi)$ be the rotation matrix of rotation of $\mathbb{R}^3$ about the e_1-axis through φ and let E be the reflection matrix at the e_1-e_2-plane, i.e.

$$e_1 = \begin{pmatrix}1\\0\\0\end{pmatrix}, \qquad e_2 = \begin{pmatrix}0\\1\\0\end{pmatrix}, \qquad e_3 = \begin{pmatrix}0\\0\\1\end{pmatrix},$$

$$R(\varphi) = \begin{pmatrix}1 & 0 & 0\\ 0 & \cos(\varphi) & -\sin(\varphi)\\ 0 & \sin(\varphi) & \cos(\varphi)\end{pmatrix}, \qquad E = \begin{pmatrix}1 & 0 & 0\\ 0 & 1 & 0\\ 0 & 0 & -1\end{pmatrix}.$$

Let furthermore $t_i = (1, e_i)$, $i \in \{1,2,3\}$ and I be the identity map. Then the Bieberbach groups can be described as follows (see e.g. [KK03]):

	generators of G_i
G_1	t_1, t_2, t_3
G_2	t_1, t_2, t_3 and $\alpha = (R_\pi, \frac{1}{2}e_1)$
G_3	$t_1, s_1 = (I, R_{\frac{2\pi}{3}}e_2), s_2 = (I, (R_{\frac{4\pi}{2}}e_2))$ and $\alpha = (R_{\frac{2\pi}{3}}, \frac{1}{3}e_1)$
G_4	t_1, t_2, t_3 and $\alpha = (R_{\frac{\pi}{2}}, \frac{1}{4}e_1)$
G_5	$t_1, s_1 = (I, R_{\frac{\pi}{3}}e_2), s_2 = (R(\frac{2\pi}{3})e_2, I)$ and $\alpha = (R_{\frac{\pi}{3}}, \frac{1}{6}e_1)$
G_6	$t_1, t_2, t_3, \alpha = (R_\pi, \frac{1}{2}e_1)$, $\beta = (-E \cdot R_\pi, \frac{1}{2}(e_2 + e_3))$ and $\gamma = (-E, \frac{1}{2}(e_1 + e_2 + e_3))$
G_7	t_1, t_2, t_3 and $\alpha = (E, \frac{1}{2}e_1)$
G_8	$t_1, t_2, s = (I, \frac{1}{2}(e_1 + e_2) + e_3)$ and $\alpha = (E, \frac{1}{2}e_1)$
G_9	$t_1, t_2, t_3, \alpha = (R_\pi, \frac{1}{2}e_1)$ and $\beta = (E, \frac{1}{2}e_2)$
G_{10}	$t_1, t_2, t_3, \alpha = (R_\pi, \frac{1}{2}e_1)$ and $\beta = (E, \frac{1}{2}(e_2 + e_3))$

The manifolds M/G_i are orientable if $1 \leq i \leq 6$ and non-orientable if $7 \leq i \leq 10$. Now we extract the generators of the holonomy and use Proposition 3.2.4 to compute the dimension of $\ker(\Delta_E|_{TT})$:

	generators of $r(G_i)$	$\dim(\ker\Delta_E\|_{TT})$
G_1	I	5
G_2	R_π	3
G_3	$R_{\frac{2\pi}{3}}$	1
G_4	$R_{\frac{\pi}{2}}$	1
G_5	$R_{\frac{\pi}{3}}$	1
G_6	$\{R_\pi, -E\cdot R_\pi, -E\}$	2
G_7	E	3
G_8	E	3
G_9	$\{R_\pi, E\}$	2
G_{10}	$\{R_\pi, E\}$	2

This table in particular shows that each three-dimensional Bieberbach manifold has infinitesimal Einstein deformations and hence, it is also deformable as an Einstein space by our remark above. In fact, the moduli space of Einstein structures on these manifolds concides with the moduli space of flat structures . An explicit desciption of these moduli spaces is given in [Kan06, Theorem 4.5].

It seems possible but it is not known if there are Bieberbach manifolds which are isolated as Einstein spaces.

3.3 Product Manifolds

Let (M, g_1) and (N, g_2) be Einstein manifolds and consider the product manifold $(M \times N, g_1 + g_2)$. It is Einstein if and only if the components have the same Einstein constant μ. In this case, the Einstein constant of the product is also μ. We want to determine if a product Einstein space is stable or not. This was worked out in [AM11] in the case, where the Einstein constant is negative. We now study the general case.

In the following, we often lift tensors on the factors M, N to tensors on $M \times N$ by pulling back along the projecton maps. In order to avoid notational complications, we drop the explicit reference to the projections throughout the section.

At first, we consider the spectrum of the Einstein operator on the product space.

Proposition 3.3.1 ([AM11]). *Let $\Delta_E^{M\times N}$ be the Einstein operator with respect to the product metric acting on $\Gamma(S^2(M \times N))$. Then the spectrum of $\Delta_E^{M\times N}$ is given by*

$$\begin{aligned}\mathrm{spec}(\Delta_E^{M\times N}) = (\mathrm{spec}(\Delta_E^M) + \mathrm{spec}(\Delta_0^N)) &\cup (\mathrm{spec}(\Delta_E^N) + \mathrm{spec}(\Delta_0^M)) \\ &\cup (\mathrm{spec}(\Delta_1^M) + \mathrm{spec}(\Delta_1^N)).\end{aligned}$$

Here, Δ_0^M, Δ_0^N, Δ_1^M, Δ_1^N denote the connection Laplacians on functions and 1-forms with respect to the metrics on M and N, respectively.

Proof. Let $\{\alpha_i\}$, $\{\omega_i\}$, $\{h_i\}$ be complete orthonormal systems of symmetric $(0,p)$-eigentensors ($p = 0, 1, 2$) of the operators Δ_0^M, Δ_1^M, Δ_E^M, respectively. Let $\lambda_i^{(0)}, \lambda_i^{(1)}, \lambda_i^{(2)}$ be the corresponding eigenvalues. Let $\{\beta_i\}$, $\{\phi_i\}$, $\{k_i\}$ be complete orthonormal systems of symmetric $(0,p)$-eigentensors ($p = 0, 1, 2$) of the operators Δ_0^N, Δ_1^N, Δ_E^N, respectively. Let $\kappa_i^{(0)}, \kappa_i^{(1)}, \kappa_i^{(2)}$ be their eigenvalues.

By [AM11, Lemma 3.1], the tensor products $\alpha_i k_j$, $\beta_i h_j$, $\omega_i \odot \phi_j$ form a complete orthonormal system in $\Gamma(S^2(M \times N))$. Straightforward calculations show that

$$\begin{aligned}
\Delta_E^{M\times N}(\alpha_i k_j) &= (\lambda_i^{(0)} + \kappa_j^{(2)})\alpha_i k_j,\\
\Delta_E^{M\times N}(\omega_i \odot \phi_j) &= (\lambda_i^{(1)} + \kappa_j^{(1)})\omega_i \odot \phi_j,\\
\Delta_E^{M\times N}(\beta_i h_j) &= (\kappa_i^{(0)} + \lambda_j^{(2)})\beta_i h_j,
\end{aligned}$$

from which the assertion follows. □

Lemma 3.3.2. *Let (M,g) be an Einstein manifold with constant μ. Then the spectrum of Δ_E on $\Gamma(S^2M)$ can be decomposed as*

$$\mathrm{spec}(\Delta_E) = \mathrm{spec}(\Delta_0 - 2\mu\cdot \mathrm{id}) \cup \mathrm{spec}_+((\Delta_1 - \mu\cdot\mathrm{id})|_W) \cup \mathrm{spec}(\Delta_E|_{TT})$$

where $W = \{\omega \in \Omega^1(M) \mid \delta\omega = 0\}$.

Proof. If (M,g) is not the standard sphere, we consider the decomposition

$$\Gamma(S^2M) = C^\infty(M)\cdot g \oplus \delta_g^*(\Omega^1(M)) \oplus TT_g.$$

Let $\{f_i\}$, $i \in \mathbb{N}_0$ be an eigenbasis of Δ_0 to the eigenvalues $\lambda_i^{(0)}$, where f_0 is the constant eigenfunction. Let $\{\omega_i\}$, $i \in \mathbb{N}$, be an eigenbasis of $\Delta_1 = \Delta_H - \mu$ acting on W with eigenvalues $\lambda_i^{(1)}$. Let $\{h_i\}_{i\in\mathbb{N}}$ be an eigenbasis of $\Delta_E|_{TT}$ with eigenvalues $\lambda_i^{(2)}$. Then $\{\nabla f_i\}$, $i \in \mathbb{N}$, $\{\omega_i\}$, $i \in \mathbb{N}$ form an eigenbasis of Δ_1 on all 1-forms and $\{f_i \cdot g\}$, $i \in \mathbb{N}_0$, $\{\nabla^2 f_i\}$, $i \in \mathbb{N}$, $\{\delta^*\omega_i\}$, $i \in \mathbb{N}$ and $\{h_i\}$, $i \in \mathbb{N}$ form a basis of $\Gamma(S^2M)$.

If $(M,g) = (S^n, g_{sp})$, we obtain a basis, if we remove from $\{\nabla^2 f_i\}$ the f_i which are the eigenfunctions to the first nonzero eigenvalue of the Laplacian (c.f. Remark 2.4.2). By the relation $\Delta_E = \Delta_L - 2\mu\cdot\mathrm{id}$ and Lemma 2.4.5, we have

$$\begin{aligned}
\Delta_E(f_i\cdot g) &= (\lambda_i^{(0)} - 2\mu) f_i \cdot g,\\
\Delta_E(\nabla^2 f_i) &= (\lambda_i^{(0)} - 2\mu)\nabla^2 f_i,\\
\Delta_E(\delta^*\omega_i) &= (\lambda_i^{(1)} - \mu)\delta^*\omega_i,
\end{aligned}$$

which shows that we have obtained a basis of eigentensors of Δ_E. By Lemma 3.3.3 below, $\lambda_i^{(1)} - \mu \geq 0$ and equality holds if and only if $\delta^*\omega_i = 0$. This finishes the proof of the lemma. □

Lemma 3.3.3. *Let (M,g) be an Einstein manifold with constant μ and W as in Lemma 3.3.2 above. Then*

$$\|\nabla\omega\|_{L^2}^2 = 2\|\delta^*\omega\|^2 + \mu\|\omega\|_{L^2}^2$$

for any $\omega \in W$. In particular, $\mathrm{spec}((\Delta_1 - \mu\cdot\mathrm{id})|_W)$ *is nonnegative.*

Proof. Let $\{e_1, \ldots, e_n\}$ be a local orthonormal frame. Then

$$\begin{aligned}
\|\nabla\omega\|_{L^2}^2 &= \int_M \sum_{i,j} (\nabla_{e_i}\omega(e_j))^2 \, dV \\
&= \frac{1}{2}\sum_{i,j}\int_M [(\nabla_{e_i}\omega(e_j) + \nabla_{e_j}\omega(e_i))^2 - 2(\nabla_{e_i}\omega(e_j)\nabla_{e_j}\omega(e_i))] \, dV \\
&= 2\|\delta^*\omega\|^2 + \int_M \sum_{i,j} \omega(e_j)\nabla^2_{e_i,e_j}\omega(e_i) \, dV \\
&= 2\|\delta^*\omega\|^2 + \int_M \sum_{i,j} \omega(e_j) R_{e_i,e_j}\omega(e_i) \, dV \\
&= 2\|\delta^*\omega\|^2 + \int_M \sum_{j} \omega(e_j)(\omega \circ \mathrm{Ric})(e_j) \, dV \\
&= 2\|\delta^*\omega\|^2 + \mu\|\omega\|_{L^2}^2
\end{aligned}$$

and if μ is nonnegative, the nonnegativity of $\Delta_1 - \mu \cdot \mathrm{id} = \nabla^*\nabla - \mu \cdot \mathrm{id}$ follows. If μ is negative, $\Delta_1 - \mu \cdot \mathrm{id}$ is obviously positive. □

Proposition 3.3.4. *If (M, g_1) and (N, g_2) are two stable Einstein metrics with $\mu \leq 0$, the product manifold $(M \times N, g + h)$ is also stable.*

Proof. By Lemma 3.3.2 and since $\mu \leq 0$, the operators Δ_E^M, Δ_E^N are nonnegative on all of $\Gamma(S^2M)$ if and only if their restriction to TT-tensors is, respectively. By Proposition 3.3.1, $\Delta_E^{M\times N}$ is nonnegative since the sum of the spectra does not contain negative elements. □

If (M, g) and (N, g_2) are stable Einstein manifolds with constant $\mu < 0$, it is also quite immediate that

$$\ker(\Delta_E^{M\times N}|_{TT}) \cong \ker(\Delta_E^M|_{TT}) \oplus \ker(\Delta_E^N|_{TT})$$

(see [AM11, Lemma 3.2]). We show that if $\mu = 0$, the situation is slightly more subtle.

Proposition 3.3.5. *Let (M^{n_1}, g_1) and (N^{n_2}, g_2) be stable Ricci-flat manifolds. Then*

$$\begin{aligned}
\ker(\Delta_E^{M\times N}|_{TT}) \cong& \mathbb{R}(n_2 \cdot g_1 - n_1 \cdot g_2) \oplus (\mathrm{par}(M) \odot \mathrm{par}(N)) \\
&\oplus \ker(\Delta_E^M|_{TT}) \oplus \ker(\Delta_E^N|_{TT}).
\end{aligned}$$

Here, $\mathrm{par}(M), \mathrm{par}(N)$ *denote the spaces of parallel* 1*-forms on* M, N *respectively. If all infinitesimal Einstein deformations of* M *and* N *are integrable, then all infinitesimal Einstein deformations of* $M \times N$ *are integrable.*

Proof. By the proof of Proposition 3.3.1, the kernel of $\Delta_E^{M\times N}$ is spanned by tensors of the form $\alpha_i k_j$, $\beta_i h_j$, $\omega_i \odot \phi_j$ where α_i, ω_i, h_i and β_i, ϕ_i, k_i are eigentensors of $\Delta_0, \Delta_1, \Delta_E$ on M and N, respectively. By Lemma 3.3.2, these operators are nonnegative, so the eigentensors have to lie in the kernel of the corresponding operators. Moreover,

$$\ker(\Delta_E^M) = \mathbb{R} \cdot g_1 \oplus \ker(\Delta_E^M|_{TT})$$

and

$$\ker(\Delta_E^N) = \mathbb{R} \cdot g_2 \oplus \ker(\Delta_E^N|_{TT}).$$

This shows

$$\begin{aligned}\ker(\Delta_E^{M\times N}) \cong& \mathbb{R} \cdot g_1 \oplus \mathbb{R} \cdot g_2 \oplus (\mathrm{par}(M) \odot \mathrm{par}(N)) \\ &\oplus \ker(\Delta_E^M|_{TT}) \oplus \ker(\Delta_E^N|_{TT}).\end{aligned}$$

The first assertion follows from restricting $\Delta_E^{M\times N}$ to TT-tensors. Any deformation $h \in \mathbb{R}(n_2 \cdot g_1 - n_1 \cdot g_2)$ is integrable since it can be integrated to a curve of metrics of the form $(g_1)_t + (g_2)_t$ where $(g_1)_t$ and $(g_2)_t$ are just rescalings of g_1 and g_2. This of course does not affect the Ricci-flatness of $M \times N$.

Now, consider the situation where $h \in (\mathrm{par}(M) \odot \mathrm{par}(N))$. Let $\omega_1, \ldots, \omega_{m_1}$ be a basis of $\mathrm{par}(M)$ and $\phi_1, \ldots \phi_{m_2}$ be a basis of $\mathrm{par}(N)$. Suppose for simplicity that all these forms have constant lengh 1. Then

$$h = \sum_{i=1}^{m_1} \sum_{j=1}^{m_2} \alpha_{ij} \omega_i \odot \phi_j.$$

We show that h is integrable. By the holonomy principle, we have parallel decompositions

$$TM = E \oplus \bigoplus_{i=1}^{m_1} (\mathbb{R} \cdot \omega_i^\sharp), \qquad TN = F \oplus \bigoplus_{j=1}^{m_2} (\mathbb{R} \cdot \phi_j^\sharp),$$

and the metrics split as $g_1 = \tilde{g}_1 + \sum_{i=1}^{m_1} \omega_i \otimes \omega_i$, $g_2 = \tilde{g}_2 + \sum_{j=1}^{m_2} \phi_j \otimes \phi_j$. The metrics $\tilde{g}_1$ and $\tilde{g}_2$ are also Ricci-flat. The tangent bundle of the product manifold obviously splits as

$$T(M \times N) = E \oplus F \oplus \bigoplus_{i=1}^{m_1} (\mathbb{R} \cdot \omega_i^\sharp) \oplus \bigoplus_{j=1}^{m_2} (\mathbb{R} \cdot \phi_j^\sharp).$$

Observe that $g_1 + g_2$ is flat when restricted to

$$G = \bigoplus_{i=1}^{m_1} (\mathbb{R} \cdot \omega_i^\sharp) \oplus \bigoplus_{j=1}^{m_2} (\mathbb{R} \cdot \phi_j^\sharp).$$

Consider the curve of metrics $t \mapsto g_t = g_1 + g_2 + th$ on $M \times N$.

The metric restricted $E \oplus F$ does not change and stays flat if we restrict to G. Thus, g_t is a curve of Ricci-flat metrics, so h is integrable.

If $h \in \ker(\Delta_E^M|_{TT})$, then there exists a curve of Einstein metrics $(g_1)_t$ on M tangent to h by assumption. Consequently, the curve $(g_1)_t \oplus g_2$ is a curve of Einstein metrics on $M \times N$ tangent to h, so h is integrable (considered as an infinitesimal Einstein deformation on $M \times N$). If $h \in \ker(\Delta_E^N|_{TT})$, an analogous argument shows the integrability of h. □

Now, let us turn to the case where the Einstein constant is positive.

Lemma 3.3.6. *Let (M, g) be a positive Einstein manifold with constant μ. Then*

$$\dim(\ker\Delta_E) = 2 \cdot \mathrm{mult}_{\Delta_0}(2\mu) + \dim(\ker\Delta_E|_{TT}),$$
$$\mathrm{ind}(\Delta_E) = 1 + \mathrm{mult}_{\Delta_0}\left(\frac{n}{n-1}\mu\right) + \sum_{\lambda\in(\frac{n}{n-1}\mu,2\mu)} 2 \cdot \mathrm{mult}_{\Delta_0}(\lambda) + \mathrm{ind}(\Delta_E|_{TT}),$$

where $\mathrm{mult}_{\Delta_0}(\lambda)$ *is the multiplicity of* λ *as an eigenvalue of* Δ_0 *and* $\mathrm{ind}(\Delta_E)$ *is the index of the quadratic form* $h \mapsto (\Delta_E h, h)_{L^2}$.

Proof. This follows immediately from the proof of Lemma 3.3.2 and Obata's theorem (Theorem 2.4.3). □

Proposition 3.3.7. *Let (M^{n_1}, g_1), (N^{n_2}, g_2) be stable Einstein manifolds with constant $\mu > 0$. Then*

$$\begin{aligned}\dim(\ker\Delta_E^{M\times N}|_{TT}) =& \dim(\ker\Delta_E^M|_{TT}) + \dim(\ker\Delta_E^N|_{TT})\\ &+ \mathrm{mult}_{\Delta_0^M}(2\mu) + \mathrm{mult}_{\Delta_0^N}(2\mu),\\ \mathrm{ind}(\Delta_E^{M\times N}|_{TT}) =& 1 + \sum_{\lambda\in(\frac{n_1}{n_1-1}\mu,2\mu)} \mathrm{mult}_{\Delta_0^M}(\lambda) + \sum_{\lambda\in(\frac{n_2}{n_2-1}\mu,2\mu)} \mathrm{mult}_{\Delta_0^N}(\lambda).\end{aligned}$$

Proof. We now prove the first assertion. By Lemma 3.3.3, Δ_1^M and Δ_1^N are positive. Thus by Proposition 3.3.1, we have to count the number of eigenvalues (with their multiplicity) $\lambda_i^{(0)} \in \mathrm{spec}(\Delta_0^M)$, $\lambda_i^{(2)} \in \mathrm{spec}(\Delta_E^M)$, $\kappa_i^{(0)} \in \mathrm{spec}(\Delta_0^N)$, $\kappa_i^{(2)} \in \mathrm{spec}(\Delta_E^N)$ such that $\lambda_i^{(0)} + \kappa_i^{(2)} = 0$ and $\lambda_i^{(2)} + \kappa_i^{(0)} = 0$. Consider the first equation. If $\lambda_i^{(0)} = \lambda_0^{(0)} = 0$, then also $\kappa_i^{(2)} = 0$ and the multiplicity of $\kappa_i^{(2)}$ is given in Lemma 3.3.6. If $\lambda_i^{(0)} > 0$, then $\kappa_i^{(2)} < 0$. By Lemma 3.3.2, Lemma 3.3.3 and since (M, g_1) is stable, $\kappa_i^{(2)} + 2\mu = \kappa_i^{(0)} \in \mathrm{spec}(\Delta_0^N)$. We thus have to find $\kappa_i^{(0)}$ such that $\lambda_i^{(0)} + \kappa_i^{(0)} = 2\mu$ for $\lambda_i^{(0)} > 0$. By Obata's eigenvalue estimate, we have a lower bound $\lambda_i^{(0)}, \kappa_i^{(0)} \geq \frac{n}{n-1}\mu$ for nonzero eigenvalues of the Laplacian. Therefore, the only situation which remains possible is that $\lambda_i^{(0)} = 2\mu$ and $\kappa_i^{(0)} = \kappa_0^{(0)} = 0$. Since eigenvalue zero has always multplicity 1, $\kappa_i^{(2)} = \kappa_0^{(0)} - 2\mu = -2\mu$ is of multiplicity 1. Now we do the same game for the equation $\lambda_i^{(2)} + \kappa_i^{(0)} = 0$. We obtain, after summing up both cases,

$$\begin{aligned}\dim(\ker\Delta_E^{M\times N}) =& \dim(\ker\Delta_E^M|_{TT}) + \dim(\ker\Delta_E^N|_{TT})\\ &+ 3\mathrm{mult}_{\Delta_0^M}(2\mu) + 3\mathrm{mult}_{\Delta_0^N}(2\mu).\end{aligned}$$

By the formula

$$\mathrm{mult}_{\Delta_0^{M\times N}}(\tau) = \sum_{\lambda+\kappa=\tau} \mathrm{mult}_{\Delta_0^M}(\lambda) \cdot \mathrm{mult}_{\Delta_0^N}(\kappa) \tag{3.2}$$

and by Obata's eigenvalue estimate,

$$\mathrm{mult}_{\Delta_0^{M\times N}}(2\mu) = \mathrm{mult}_{\Delta_0^M}(2\mu) + \mathrm{mult}_{\Delta_0^N}(2\mu).$$

From Lemma 3.3.6, we get the dimension of $\ker\Delta_E^{M\times N}|_{TT}$.

To show the second assertion, we compute the number of eigenvalues (with multiplicity) satisfiying $\lambda_i^{(0)}+\kappa_i^{(2)}<0$ or $\lambda_i^{(2)}+\kappa_i^{(0)}<0$. Consider the first inequality. If $\lambda_i^{(0)}=\lambda_0^{(0)}=0$, then $\kappa_i^{(2)}<0$ and the number of such eigenvalues (with multiplicity) is given by Lemma 3.3.6. If $\lambda_i^{(0)}>0$, then $\lambda_i^{(0)}\geq\frac{n}{n-1}\mu$ and $\kappa_i^{(2)}<-\frac{n}{n-1}\mu$. By Lemma 3.3.2, $\kappa_i^{(2)}+2\mu=\kappa_i^{(0)}\in\operatorname{spec}(\Delta_0^N)$ and $\kappa_i^{(0)}<\frac{n-2}{n-1}\mu$. By Obata's eigenvalue estimate, $\kappa_i^{(0)}=\kappa_0^{(0)}=0$ and $\kappa_i^{(2)}=-2\mu$ appears with multiplicity 1. This also implies that $\lambda_i^{(0)}<2\mu$.

Simliarly, we deal with the inequality $\lambda_i^{(2)}+\kappa_i^{(0)}<0$. Summing up over both cases, we obtain

$$\begin{aligned}\operatorname{ind}(\Delta_E^{M\times N})=&2+3\sum_{\lambda\in(\frac{n_1}{n_1-1}\mu,2\mu)}\operatorname{mult}_{\Delta_0^M}(\lambda)+3\sum_{\lambda\in(\frac{n_2}{n_2-1},2\mu)}\operatorname{mult}_{\Delta_0^N}(\lambda)\\&+2\cdot\operatorname{mult}_{\Delta_0^M}\left(\frac{n_1}{n_1-1}\mu\right)+2\cdot\operatorname{mult}_{\Delta_0^N}\left(\frac{n_2}{n_2-1}\mu\right).\end{aligned}$$

By (3.2) and by Obata's eigenvalue estimate,

$$\sum_{\lambda\in(0,2\mu)}\operatorname{mult}_{\Delta_0^{M\times N}}(\lambda)=\sum_{\lambda\in(0,2\mu)}\operatorname{mult}_{\Delta_0^M}(\lambda)+\sum_{\lambda\in(0,2\mu)}\operatorname{mult}_{\Delta_0^N}(\lambda)$$

and the second assertion follows from Lemma 3.3.6. □

As we see, products of positive Einstein manifolds are always unstable and small eigenvalues of the Laplacian enlarge the index of the form

$$TT\ni h\mapsto(\Delta_E h,h)_{L^2}.$$

Remark 3.3.8. In particular, if 2μ is an eigenvalue of the Laplace-Beltrami operator on M or N (this holds e.g. for the complex projective space) then the product metric has infinitesimal Einstein deformations. The non-integrable infinitesimal Einstein deformations on $\mathbb{C}P^{2n}\times S^2$ mentioned in Section 2.5 are of this form.

Chapter 4

Stability and Curvature

In this chapter, we study curvature conditions which ensure stability of Einstein manifolds. We build upon work by Koiso ([Koi78; Koi79b; Koi80; Koi82; Koi83]), Itoh and Nagakawa ([IN05]).

4.1 Stability under Sectional Curvature Bounds

At first, we mention an important theorem by Koiso, which is a first attempt to relate stability of Einstein manifolds to curvature assumptions. Because we also work with his methods later on, we will sketch Koiso's proof of the theorem below. Let S^2_gM be the vector bundle of symmetric $(0,2)$-tensors whose trace with respect to g vanishes.

Theorem 4.1.1 ([Koi78]). *Let (M,g) be Einstein with constant μ. Let r_0 be the largest eigenvalue of $\mathring{R}$ on traceless tensors, i.e.*

$$r_0 = \sup\left\{ \frac{(\mathring{R}h,h)_{L^2}}{\|h\|^2_{L^2}} \,\middle|\, h \in \Gamma(S^2_gM) \right\}. \tag{4.1}$$

If $r_0 \leq \max\left\{-\mu, \frac{1}{2}\mu\right\}$, then (M,g) is stable. If $r_0 < \max\left\{-\mu, \frac{1}{2}\mu\right\}$, then (M,g) is strictly stable.

Proof. We define two differential operators by

$$\begin{aligned} D_1h(X,Y,Z) &= \frac{1}{\sqrt{3}}(\nabla_X h(Y,Z) + \nabla_Y h(Z,X) + \nabla_Z h(X,Y)), \\ D_2h(X,Y,Z) &= \frac{1}{\sqrt{2}}(\nabla_X h(Y,Z) - \nabla_Y h(Z,X)). \end{aligned}$$

For the Einstein operator, we have the Bochner formulas

$$(\Delta_E h,h)_{L^2} = \|D_1h\|^2_{L^2} + 2\mu\,\|h\|^2_{L^2} - 4(\mathring{R}h,h)_{L^2} - 2\,\|\delta h\|^2_{L^2}, \tag{4.2}$$

$$(\Delta_E h,h)_{L^2} = \|D_2h\|^2_{L^2} - \mu\,\|h\|^2_{L^2} - (\mathring{R}h,h)_{L^2} + \|\delta h\|^2_{L^2}, \tag{4.3}$$

see [Koi78] or [Bes08, p. 355] for more details. Because of the bounds on r_0 and $\delta h = 0$, we obtain either $(\Delta_E h,h)_{L^2} \geq 0$ or $(\Delta_E h,h)_{L^2} > 0$ for TT-tensors by (4.2) or (4.3). □

The next step is to estimate r_0 in terms of sectional curvature bounds. We define a function on M by

$$r(p) = \sup\left\{ \frac{\langle \mathring{R}\eta, \eta\rangle_p}{|\eta|_p^2} \,\middle|\, \eta \in (S_g^2 M)_p \right\}. \tag{4.4}$$

Observe that $r_0 \leq \sup_{p\in M} r(p)$.

Lemma 4.1.2 ([Fuj79], see also [Bes08]). *Let (M,g) be Einstein and $p \in M$. Let K_{min} and K_{max} be the minimum and maximum of its sectional curvature at p, then*

$$r(p) \leq \min\{(n-2)K_{max} - \mu, \mu - nK_{min}\}.$$

Proof. Choose η such that $\mathring{R}\eta = r(p)\eta$. Let $\{e_1, \ldots, e_n\}$ be an orthonormal basis in which η is diagonal with eigenvalues $\lambda_1, \ldots, \lambda_n$ such that $\lambda_1 = \sup|\lambda_i|$ and $\sum \lambda_i = 0$. Then

$$r(p)\lambda_1 = (\mathring{R}\eta)(e_1,e_1) = \sum_{i,j} R(e_i,e_1,e_1,e_j)h(e_j,e_i) = \sum_i K_{i1}\lambda_i$$

where K_{i1} is the sectional curvature of the plane spanned by e_i and e_1. Thus,

$$\begin{aligned} r(p)\lambda_1 &= \sum_{i\neq 1} K_{max}\lambda_i - \sum_{i\neq 1}(K_{max} - K_{i1})\lambda_i \\ &\leq -\lambda_1 K_{max} + \lambda_1 \sum_{i\neq 1}(K_{max} - K_{i1}) \\ &= ((n-2)K_{max} - \mu)\lambda_1. \end{aligned} \tag{4.5}$$

On the other hand,

$$\begin{aligned} r(p)\lambda_1 &= \sum_{i\neq 1} K_{min}\lambda_i + \sum_{i\neq 1}(K_{i1} - K_{min})\lambda_i \\ &\leq -\lambda_1 K_{min} + \lambda_1 \sum_{i\neq 1}(K_{i1} - K_{min}) \\ &= (-nK_{min} + \mu)\lambda \end{aligned} \tag{4.6}$$

which finishes the proof. □

As a consequence, we get two well-known corrollaries.

Corollary 4.1.3. *Any Einstein manifold (M,g) with $\frac{n-2}{3n}$-pinched sectional curvature, i.e., its sectional cuvature lies in the half-open interval $(\frac{n-2}{3n}, 1]\cdot K_{max}$, is strictly stable.*

Proof. This assumption means that $2nK_{min} > \frac{2}{3}(n-2)K_{max}$. Therefore, either $\mu > \frac{2}{3}(n-2)K_{max}$ or $\mu < 2nK_{min}$. In both cases, $r_0 \leq \sup r(p) < \frac{\mu}{2}$ by Lemma 4.1.2 and Theorem 4.1.1 implies strict stability. □

Corollary 4.1.4. *Any Einstein manifold (M,g) of negative sectional curvature is strictly stable.*

Proof. By Lemma 4.1.2, $K_{max} < 0$ implies $r_0 \leq \sup r(p) < -\mu$ and strict stability again follows from Theorem 4.1.1. □

By the last corollary, we obtained a quite strong stability criterion for negative Einstein metrics. However, the pinching criterion from Corollary 4.1.3 is rather weak, because it is only of use in dimension $n < 8$. For $n \geq 8$, any Einstein manifold satisfying the curvature conditions of Corollary 4.1.3 is quater-pinched and thus, by the proof of the differentiable sphere theorem, it is isometric to a quotient the standard sphere (see [BS09]).

4.2 Extensions of Koiso's Results

Now, we want to prove stability under weaker conditions than in the Corollaries 4.1.3 and 4.1.4. Unfortunately we cannot go further than replacing the strict inequalities in the assumptions by weak inequalities. Then we immediately get $\Delta_E|_{TT} \geq 0$. Furthermore, we will see that the existence of infinitesimal Einstein deformations imposes very strict conditions on the structure of the manifold. We first need a technical lemma.

Lemma 4.2.1. *Let (M^n, g) be Einstein with constant μ and $p \in M$. Suppose that*

$$r(p) = (n-2)K_{max} - \mu = \mu - nK_{min}. \tag{4.7}$$

Here, $r(p)$ is the function defined in (4.4) and K_{max}, K_{min} are the maximal and minimal sectional curvatures of planes lying in T_pM, respectively.

Then (M, g) is even-dimensional. Let $\eta \in (S^2_gM)_p$ be such that $\mathring{R}\eta = r(p)\eta$. Then η has only two eigenvalues $\lambda, -\lambda$ and the eigenspaces $E(\lambda), E(-\lambda)$ are both of dimension $m = n/2$. Moreover, $K(P) = K_{max}$ for each plane P lying in either $E(\lambda)$ or $E(-\lambda)$ and $K(P) = K_{min}$ if P is spanned by one vector in $E(\lambda)$ and one in $E(-\lambda)$.

Proof. Let $\eta \in (S^2_gM)_p$ be such that $\mathring{R}h = r(p)h$. As in the proof of Lemma 4.1.2, let $\lambda_1, \ldots, \lambda_n$ be the eigenvalues of η (with $\lambda_1 = \max|\lambda_i|$) and let K_{ij} be the sectional curvatures with respect to the corresponding orthonormal basis. By (4.7), we see that equality must hold both in (4.5) and (4.6), i.e.

$$-\sum_{i\neq 1}(K_{max} - K_{i1})\lambda_i = \lambda_1 \sum_{i\neq 1}(K_{max} - K_{i1}) \tag{4.8}$$

and

$$\sum_{i\neq 1}(K_{i1} - K_{min})\lambda_i = \lambda_1 \sum_{i\neq 1}(K_{i1} - K_{min}). \tag{4.9}$$

From (4.8), we get that either $\lambda_i = -\lambda_1$ or $K_{ij} = K_{max}$ whereas (4.9) implies $\lambda_i = \lambda_1$ or $K_{ij} = K_{min}$ for each i. Thus there only exist two eigenvalues λ and $-\lambda$ which are of same multiplicity since the trace of η vanishes. In particular, (M, g) is even-dimensional.

Let $P \subset T_pM$ be a plane which satisfies one of the assumptions of the lemma. We then may assume that P is spanned by two vectors of the eigenbasis we have chosen. If $P \subset E(\lambda)$ or P is spanned by two vectors in $E(\lambda)$, $E(-\lambda)$, respectively, we may assume $e_1 \in P$. Then the assertions follow from the above. If $P \subset E(-\lambda)$, we may replace η by $-\eta$ and the roles of $E(\lambda)$ and $E(-\lambda)$ interchange. □

Now we are able to improve Corollary 4.1.3 by considering the case where the manifold is weakly $(n-2)/3n$-pinched.

Proposition 4.2.2. *Let (M,g) be an Einstein manifold such that the sectional curvature lies in the interval $[(n-2)/3n, 1] \cdot K_{max}$, $K_{max} > 0$. Then (M,g) is stable. If $\ker \Delta_E|_{TT}$ is nontrivial, M^n is even-dimensional. Furthermore, there exists an orthogonal splitting $TM = \mathcal{E} \oplus \mathcal{F}$ into two subbundles of dimension $n/2$. The two $C^\infty(M)$-bilinear maps*

$$I\colon \Gamma(\mathcal{E}) \times \Gamma(\mathcal{E}) \to \Gamma(\mathcal{F}), \qquad (X,Y) \mapsto pr_{\mathcal{F}}(\nabla_X Y)$$

and

$$II\colon \Gamma(\mathcal{F}) \times \Gamma(\mathcal{F}) \to \Gamma(\mathcal{E}), \qquad (X,Y) \mapsto pr_{\mathcal{E}}(\nabla_X Y)$$

are both antisymmetric in X and Y. Moreover, the sectional curvature of a plane P is equal to K_{max} if P either lies in $\mathcal{E}$ or $\mathcal{F}$. If $P = \mathrm{span}\{e, f\}$ with $e \in \mathcal{E}$ and $f \in \mathcal{F}$, then $K(P) = K_{min}$.

Proof. Let μ be the Einstein constant. Because of curvature assumpions, $\mu \geq \frac{2}{3}(n-2)K_{max}$ or $\mu \leq 2nK_{min}$ at each point. In both cases, the function r from Lemma 4.1.2 satisfies $r \leq \frac{1}{2}\mu$. Thus, $r_0 \leq \frac{1}{2}\mu$

and Theorem 4.1.1 implies that (M,g) is stable. Suppose now there exists $h \in \ker\Delta_E|_{TT}$, $h \neq 0$. Then by (4.3),

$$\begin{aligned} 0 = (\Delta_E h, h)_{L^2} &= \|D_1 h\|_{L^2}^2 + 2\mu \|h\|_{L^2}^2 - 4(h, \mathring{R}h)_{L^2} \\ &\geq 0 + 2\mu \|h\|_{L^2}^2 - 2\mu \|h\|_{L^2}^2 = 0. \end{aligned}$$

Therefore, $D_1 h \equiv 0$ and $\langle \mathring{R}h, h\rangle_p \equiv \frac{\mu}{2}|h|_p^2$ for all $p \in M$. The second equality implies that

$$\mu = \frac{2}{3}(n-2)K_{max} = 2nK_{min}$$

and

$$r(p) = (n-2)K_{max} - \mu = \mu - nK_{min}.$$

Thus, Lemma 4.2.1 applies and at each point where $h \neq 0$, the tangent space splits into the two eigenspaces of h, i.e. $T_pM = E_p(\lambda) \oplus E_p(-\lambda)$. Since $D_1 h \equiv 0$, we have

$$\nabla_{e_i} h(e_j, e_k) + \nabla_{e_j} h(e_k, e_i) + \nabla_{e_k} h(e_i, e_j) = 0$$

for any local orthonormal frame $\{e_1, \dots, e_n\}$. By considering h as an endomorphism $h: TM \to TM$,

$$g(\nabla_{e_i} h(e_j), e_k) + g(\nabla_{e_j} h(e_k), e_i) + g(\nabla_{e_k} h(e_i), e_j) = 0 \tag{4.10}$$

for $1 \leq i,j,k \leq n$. Choose an eigenframe of h around some p outside the zero set of h. We compute

$$\begin{aligned}
\langle \nabla_{e_i} h(e_j), e_k \rangle &= \langle \nabla_{e_i}(h(e_j)), e_k \rangle - \langle h(\nabla_{e_i} e_j), e_k \rangle \\
&= \langle \nabla_{e_i}(\lambda_j e_j), e_k \rangle - \sum_l \langle h(\Gamma_{ij}^l e_l), e_k \rangle \\
&= \langle (\nabla_{e_i} \lambda_j) e_j, e_k \rangle + \langle \lambda_j \nabla_{e_i} e_j, e_k \rangle - \sum_l \langle \Gamma_{ij}^l \lambda_l e_l, e_k \rangle \\
&= (\nabla_{e_i} \lambda_j) \delta_{jk} + \lambda_j \Gamma_{ij}^k - \lambda_k \Gamma_{ij}^k,
\end{aligned}$$

where λ_j is the eigenvalue of e_j. Now we rewrite (4.10) as

$$\begin{aligned}
(\lambda_j - \lambda_k)\Gamma_{ij}^k + (\lambda_k - \lambda_i)\Gamma_{jk}^i + (\lambda_i - \lambda_j)\Gamma_{ki}^j& \\
= -(\nabla_{e_i}\lambda_j)\delta_{jk} - (\nabla_{e_j}\lambda_k)\delta_{ki} - (\nabla_{e_k}\lambda_i)\delta_{ij}&.
\end{aligned} \tag{4.11}$$

If we choose $i = j = k$, we obtain

$$0 = -3(\nabla_{e_i}\lambda_i).$$

Since $\lambda_i = \pm\lambda$, it is immediate that λ is constant and it is nonzero. Thus, we obtain a global splitting $TM = \mathcal{E} \oplus \mathcal{F}$ where the two distributions are defined by

$$\mathcal{E} = \bigcup_{p \in M} E_p(\lambda), \qquad \mathcal{F} = \bigcup_{p \in M} E_p(-\lambda).$$

By Lemma 4.2.1, the assertion about the sectional curvatures is immediate. To finish the proof, it just remains to show the antisymmetry of the maps I, II, respectively.

Let $\{e_1, \ldots, e_n\}$ be the eigenframe from before. Suppose that $e_1, \ldots, e_{n/2}$ are local sections in $\mathcal{E}$ and $e_{n/2+1}, \ldots, e_n$ are local sections in $\mathcal{F}$. Choose $i, j \in \{1, \ldots, n/2\}$, $k \in \{n/2+1, \ldots, n\}$. Then $\lambda_i = \lambda_j = \lambda$, $\lambda_k = -\lambda$ and (4.11) yields

$$0 = 2\lambda\Gamma_{ij}^k - 2\lambda\Gamma_{jk}^i = 2\lambda(\Gamma_{ij}^k + \Gamma_{ji}^k), \tag{4.12}$$

since the right-hand side of (4.11) vanishes for any i, j, k. Now consider the map I. It is easy to check that I is $C^\infty(M)$-bilinear in both variables. We have

$$I(e_i, e_j) = \mathrm{pr}_{\mathcal{F}}(\nabla_{e_i} e_j) = \mathrm{pr}_{\mathcal{F}}\left(\sum_{k=1}^n \Gamma_{ij}^k e_k\right) = \sum_{k=n/2+1}^n \Gamma_{ij}^k e_k, \tag{4.13}$$

and by (4.12), we immediately get $I(e_i, e_j) = -I(e_j, e_i)$. Similarly, antisymmetry is shown for II. □

Now let us turn to the case of nonpositive secional curvature.

Definition 4.2.3. Let (M, g) be a Riemannian manifold and let $\{e_1, \ldots, e_n\}$ be an orthonormal frame at $p \in M$. Then $K_{ij} = R_{ijji}$ is the sectional curvature of the plane spanned by e_i and e_j if $i \neq j$ and is zero if $i = j$. We count the number of j such that $K_{i_0 j} = 0$ for a given i_0 and call the maximum of such numbers over all orthonormal frames at p the flat dimension of M at p, denoted by $\mathrm{fd}(M)_p$. The number $\mathrm{fd}(M) = \sup_{p \in M} \mathrm{fd}(M)_p$ is called the flat dimension of M.

Proposition 4.2.4 ([Koi78]). *Let (M,g) be a non-flat Einstein manifold with nonpositive sectional curvature. Then (M,g) is stable. If $\ker(\Delta_E|_{TT})$ is nontrivial, the flat dimension of M satisfies $fd(M)_p \geq \lceil \frac{n}{2} \rceil$ at each $p \in M$.*

If in addition, a lower bound on the sectional curvature is assumed, we obtain stronger consequences of the existence of infinitesimal Einstein deformations:

Proposition 4.2.5. *Let (M,g) a non-flat Einstein manifold with nonpositive sectional curvature and Einstein constant μ. If $K_{min} > \frac{2}{n}\mu$, then (M,g) is strictly stable. If $K_{min} \geq \frac{2}{n}\mu$, then (M,g) is stable. If $\ker\Delta_E|_{TT}$ is nontrivial, then M is even-dimensional and we have an orthogonal splitting $TM = \mathcal{E} \oplus \mathcal{F}$. Both subbundles are of dimension $n/2$. The $C^\infty(M)$-bilinear maps*

$$I\colon \Gamma(\mathcal{E}) \times \Gamma(\mathcal{E}) \to \Gamma(\mathcal{F}), \qquad (X,Y) \mapsto pr_{\mathcal{F}}(\nabla_X Y)$$

and

$$II\colon \Gamma(\mathcal{F}) \times \Gamma(\mathcal{F}) \to \Gamma(\mathcal{E}), \qquad (X,Y) \mapsto pr_{\mathcal{E}}(\nabla_X Y)$$

are symmetric. Moreover, $K(P) = 0$ for any plane lying in $\mathcal{E}$ or $\mathcal{F}$.

Proof. Since the sectional curvature is nonpositive but not identically zero, the Einstein constant is negative. Now we follow the same strategy as in the proof of Proposition 4.2.2. If $K_{min} > \frac{2}{n}\mu$,then $r_p < -\mu$ and by Proposition 4.1.1, (M,g) is strictly stable. If $K_{min} \geq \frac{2}{n}\mu$ and $h \in \ker(\Delta_E|_{TT})$, we obtain from (4.3) that

$$\begin{aligned} 0 = (\Delta_E h, h)_{L^2} &= \|D_2 h\|^2_{L^2} - \mu\,\|h\|^2_{L^2} - (h, \mathring{R}h)_{L^2} \\ &\geq -\mu\,\|h\|^2_{L^2} + \mu\,\|h\|^2_{L^2} = 0. \end{aligned}$$

Consequently, $D_2 h \equiv 0$ and $r(p) = K_{max} - \mu = \mu - nK_{min}$. Again by Lemma 4.2.1, there is a splitting $T_pM = E_p(\lambda) \oplus E_p(-\lambda)$ at each point $p \in M$ where $h \neq 0$ and $E_p(\pm\lambda)$ is the $n/2$-dimensional eigenspaces of h to the eigenvalue $\pm\lambda$, respectively. Evidently, (M,g) is even-dimensional. We will now show that λ is constant in p. Let $\{e_1,\ldots,e_n\}$ be a local eigenframe of h such that $e_1,\ldots,e_{n/2} \in E(\lambda)$ and $e_{n/2+1},\ldots,e_n \in E(-\lambda)$ and let $\lambda_1 \equiv \ldots \equiv \lambda_{n/2}$ and $\lambda_{n/2+1} \equiv \ldots \equiv \lambda_n$ be the corresponding eigenfunctions. Since $D_2 h \equiv 0$,

$$(\lambda_j - \lambda_k)\Gamma^k_{ij} - (\lambda_i - \lambda_k)\Gamma^k_{ij} = -(\nabla_{e_i}\lambda_j)\delta_{jk} + (\nabla_{e_j}\lambda_i)\delta_{ik} \tag{4.14}$$

for $1 \leq i,j,k \leq n$. Choose $i \neq j$ and $j = k$ such that e_i, e_j, e_k lie in the same eigenspace. Then by (4.14),

$$0 = -\nabla_{e_i}\lambda_j$$

and since λ_j equals either λ or $-\lambda$, the eigenvalues of h are constant in p. A splitting of the tangent bundle is obtained by $TM = \mathcal{E} \oplus \mathcal{F}$ where the two distributions are defined by

$$\mathcal{E} = \bigcup_{p\in M} E_p(\lambda), \qquad \mathcal{F} = \bigcup_{p\in M} E_p(-\lambda).$$

The flatness of planes in $\mathcal{E}$ and $\mathcal{F}$ follows from Lemma 4.2.1. It remains to show the symmetry of I and II. Let $\{e_1,\ldots e_n\}$ an orthonormal frame such that $e_1, \dots, e_{n/2}$ are local sections in $\mathcal{E}$ and $e_{n/2+1}, \dots, e_n$ are local sections in $\mathcal{F}$. Let $i, j \in \{1, \dots n/2\}$ and $k \in \{n/2+1, \dots, n\}$. By (4.14),

$$2\lambda\Gamma_{ij}^k - 2\lambda\Gamma_{ji}^k = 0.$$

and since

$$I(e_i, e_j) = \sum_{k=n/2+1}^{n} \Gamma_{ij}^k e_k, \tag{4.15}$$

I is symmetric. The symmetry of II is shown by the same arguments. It is furthermore easy to see that both maps are $C^\infty(M)$-bilinear. □

Remark 4.2.6. By symmetry of the operators I and II, the map $(X, Y) \mapsto [X, Y]$ preserves the splitting $TM = \mathcal{E} \oplus \mathcal{F}$. Thus, both distributions are integrable by the Frobenius theorem.

It is not known whether the pinching assumptions of Proposition 4.2.2 can be further improved. We conclude this section with some eigenvalue estimates for the Einstein operator.

Proposition 4.2.7. *Let (M, g) be a Riemannian manifold of constant curvature K. Then (M, g) is stable. If $K \neq 0$, it is strictly stable. Let λ be the smallest eigenvalue of $\Delta_E|_{TT}$. It satisfies the estimate*

$$\lambda \geq \max\{2(n+1)K, -(n-2)K\}.$$

Proof. The stability properties of constant curvature metrics have been shown in Section 3.2 and the Corollaries 4.1.3 and 4.1.4. It remains to show the eigenvalue estimates.

For constant curvature metrics, the Riemann curvature tensor is given by $R_{X,Y}Z = K(g(Y, Z)X - g(X, Z)Y)$ and we have $\mu = (n-1)K$ for the Einstein constant. The action of the cuvature tensor on traceless tensors is given by $\mathring{R}h(X, Y) = -Kh(X, Y)$. Now, Bochner formula (4.2) yields

$$\begin{aligned}(\Delta_E h, h)_{L^2} &= \|D_1 h\|_{L^2}^2 + 2\mu \|h\|_{L^2}^2 - 4(h, \mathring{R}h)_{L^2} \\ &\geq 2(n+1)K \|h\|_{L^2}^2,\end{aligned}$$

and from (4.3), we obtain

$$\begin{aligned}(\Delta_E h, h)_{L^2} &= \|D_2 h\|_{L^2}^2 - \mu \|h\|_{L^2}^2 - (h, \mathring{R}h)_{L^2} \\ &\geq -(n-2)K \|h\|_{L^2}^2.\end{aligned}$$

□

Remark 4.2.8. For nonnegative K, this lower bound is optimal. It is achieved on the torus and the sphere, see Examples 3.1.1 and 3.1.2. For hyperbolic spaces, this is not known but should be numerically computable.

Proposition 4.2.9. *Let (M, g) an Einstein manifold with constant μ and sectional curvature $K \geq 0$. Then the smallest eigenvalue of $\Delta_E|_{TT}$ satisfies*

$$\lambda \geq -2\mu.$$

Moreover, equality holds if and only if the holonomy of (M, g) is reducible.

Proof. By curvature assumptions and Lemma 4.1.2, $r_0 \leq \sup_{p\in M} r(p) \leq \mu$, where r_0 and $r(p)$ are defined in (4.1) and (4.4), respectively. Therefore,

$$(\Delta_E h, h) = \|\nabla h\|_{L^2}^2 - 2(\mathring{R}h, h)_{L^2} \geq -2\mu \|h\|_{L^2}^2,$$

and equality implies that h is parallel. By Lemma 3.2.2, the holonomy of (M, g) is reducible. Conversely, if (M, g) has reducible holonomy, the metric splits as $g = g_1 + g_2$ and a tracefree linear combination $\alpha g_1 + \beta g_2$ is an eigentensor of $\Delta_E|_{TT}$ to the eigenvalue -2μ. □

Remark 4.2.10. If the holonomy is reducible, (M, g) is locally isometric to a Riemannian product $(M_1, g_1) \times (M_2, g_2)$. This follows from [Bau09, Satz 5.6]. In particular, the sectional curvature cannot be positive in this case.

From the previous proposition we can deduce the following assertion for the Lichnerowicz Laplacian:

Proposition 4.2.11. *Let (M, g) be an Einstein manifold with nonnegative sectional curvature. Then the Lichnerowicz Laplacian is positive semidefinite on $\Gamma(S^2M)$ and* $\mathrm{span}(g) \subset \ker\Delta_L$. *Moreover (M, g) has reducible holonomy if and only if* $\mathrm{span}(g) \subsetneq \ker\Delta_L$,

Proof. Obviously, the Einstein constant μ is nonnegative. If (M, g) is Ricci-flat, it is flat by our curvature assumptions and the Lichnerowicz Laplacian coincides with the connection Laplacian. In this case, the assertion follows from Lemma 3.2.2. We assume $\mu > 0$ from now on. We know that Δ_L preserves on each component of the splitting

$$\Gamma(S^2M) = (C^\infty(M) \cdot g + \delta^*(\Omega^1(M))) \oplus TT.$$

By Proposition 4.2.9 and since $\Delta_L = \Delta_E + 2\mu \cdot \mathrm{id}$, Δ_L is nonnegative on TT and has nontrivial kernel if and only if (M, g) has reducible holonomy. On the first component of the splitting, Δ_L acts as the Laplace-Beltrami operator (c.f. Lemma 2.4.5) and the kernel is given by $\mathbb{R} \cdot g$. Also by Lemma 2.4.5, the spectrum of Δ_L on $\delta^*(\Omega^1(M))$ is contained in the spectrum of the Hodge Laplacian $\Delta_H = \nabla^*\nabla + \mu \cdot \mathrm{id}$ on 1-forms. Since $\mu > 0$, Δ_H is positive so Δ_L is positive on $\delta^*(\Omega^1(M))$.

□

Remark 4.2.12. Under the conditions of Proposition 4.2.9 and 4.2.11, we see that the kernel of the Lichnerowicz Laplacian consists precisely of symmetric parallel $(0, 2)$-tensors. We have computed their dimension in terms of the holonomy, see Proposition 3.2.4.

Remark 4.2.13. The nonnegativity of Δ_L under these conditions also follows from the results in [Bar93, Section 2]. A charcterization of the kernel is not given there.

4.3 Stability and Weyl Curvature

We have seen that constant curvature metrics and sufficiently pinched Einstein manifolds are stable. This motivates to prove stability theorems in terms of the Weyl tensor which measures the deviation of an Einstein manifold of being of constant curvature.

Definition 4.3.1. Let h and k be two symmetric $(0, 2)$-tensors. The Kulkarni-Nomizu product of h and k is the $(0, 4)$-tensor given by

$$\begin{aligned}(h \owedge k)(X, Y, Z, W) = h(X, W)k(Y, Z) + h(Y, Z)k(X, W) \\ - h(X, Z)k(Y, W) - h(Y, W)k(X, Z).\end{aligned}$$

Any Kulkarni-Nomizu product has the same symmetries as the Riemann tensor, i.e.

$$\begin{aligned}&(h \owedge k)(X, Y, Z, W) = -(h \owedge k)(Y, X, Z, W) = -(h \owedge k)(X, Y, W, Z),\\&(h \owedge k)(X, Y, Z, W) = (h \owedge k)(Z, W, X, Y),\\&(h \owedge k)(X, Y, Z, W) + (h \owedge k)(Y, Z, X, W) + (h \owedge k)(Z, X, Y, W) = 0.\end{aligned}$$

Recall that on a metric g with constant curvature K, the Riemann tensor is given by $R = \frac{K}{2}(g \owedge g)$. With this notation, we now can formulate the Ricci decomposition of the Riemann curvature tensor. Let $\mathrm{Ric}^0 = \mathrm{Ric} - \frac{1}{n}\mathrm{scal} \cdot g$ be the traceless part of the Ricci tensor. Then the Riemann curvature tensor (considered as a $(0, 4)$-tensor) can be decomposed as

$$R = W + \frac{\mathrm{scal}}{2n(n-1)}(g \owedge g) + \frac{1}{n-2}(\mathrm{Ric}^0 \owedge g), \tag{4.16}$$

and this composition is orthogonal in the sense that

$$|R|^2 = |W|^2 + \left|\frac{\mathrm{scal}}{2n(n-1)}(g \owedge g)\right|^2 + \left|\frac{1}{n-2}(\mathrm{Ric}^0 \owedge g)\right|^2$$

(see [Bes08, p. 48]). We call the tensor W, defined by equation (4.16), the Weyl cuvature tensor. It has the same symmetries as the Riemann curvature tensor. Moreover, any trace of W vanishes. It also has a nice behavior under conformal transformations. Suppose that the metrics $g, \tilde{g}$ are conformally equivalent, i.e. $\tilde{g} = f \cdot g$ for some smooth function $f > 0$. Then the corresponding Weyl tensors are related by $\tilde{W} = f \cdot W$, c.f. [Bes08, p. 58].

If (M, g) is Einstein with constant μ, then Ric^0 vanishes and (4.16) simplifies to

$$R = W + \frac{\mu}{2(n-1)}(g \owedge g). \tag{4.17}$$

Define $W_{ij} := W(e_i, e_j, e_j, e_i)$ for a chosen orthonormal frame. By (4.17), we have $W_{ij} = K_{ij} - \frac{\mu}{(n-1)}$ for $i \neq j$ where K_{ij} is the sectional curvature of the plane spanned by e_i and e_j. Thus, the coefficient W_{ij} measures how the sectional curvature of the plane spanned by e_i and e_j differs from its mean. The sectional curvature is constant if and only if all W_{ij} and the whole Weyl tensor vanish. We define the action of the Weyl tensor on $\Gamma(S^2M)$ by

$$\mathring{W}h(X, Y) = \sum_{i=1}^{n} h(W_{e_i, X}Y, e_i),$$

where $\{e_1, \dots, e_n\}$ is an orthonormal frame and

$$g(W_{X,Y}Z, W) = W(X, Y, Z, W).$$

By (4.17), the action of the curvature tensor on traceless tensors decomposes as

$$
\begin{aligned}
\mathring{R}h(X,Y) &= \sum_{i=1}^{n} h(R_{e_i,X}Y, e_i) \\
&= \sum_{i=1}^{n} h(W_{e_i,X}Y, e_i) + \frac{\mu}{n-1}\sum_{i=1}^{n} \{h(g(X,Y)e_i, e_i) - h(g(e_i,Y)X, e_i)\} \\
&= \sum_{i=1}^{n} h(W_{e_i,X}Y, e_i) + \frac{\mu}{n-1}\{g(X,Y)\mathrm{tr}h - h(X,Y)\} \\
&= \mathring{W}h(X,Y) - \frac{\mu}{n-1}h(X,Y).
\end{aligned}
$$

Lemma 4.3.2. *Let (M,g) be any Riemannian manifold and let $p \in M$. The operator $\mathring{W} : (S^2_gM)_p \to (S^2_gM)_p$ is trace-free. It is indefinite as long as $W_p \neq 0$.*

Proof. First we compute the trace of $\mathring{W}$ acting on all symmetric $(0,2)$-tensors. Let $\{e_1, \ldots, e_n\}$ be an orthonormal basis of T_pM. Then an orthonormal basis of $(S^2M)_p$ is given by

$$
\eta_{(ij)} = \frac{1}{\sqrt{2}} e_i^* \odot e_j^*, \qquad 1 \leq i \leq j \leq n,
$$

where $\odot$ denotes the symmetric tensor product. Simple calculations yield

$$
\langle \mathring{W}\eta_{(ij)}, \eta_{(ij)} \rangle = -W_{ijji}.
$$

Thus,

$$
\mathrm{tr}\mathring{W} = \sum_{1\leq i\leq j\leq n} \langle \mathring{W}\eta_{(ij)}, \eta_{(ij)} \rangle = -\sum_{1\leq i\leq j\leq n} W_{ijji} = -\frac{1}{2}\sum_{i,j=1}^{n} W_{ijji} = 0
$$

because the Weyl tensor has vanishing trace. Since $(\mathring{W}g)_{ij} = \sum_k W_{kijk} = 0$, the restricion of $\mathring{W}$ to $(S^2_gM)_p$ has also vanishing trace. Suppose now that the operator $\mathring{W}$ vanishes, then all W_{ijji} vanish. By the symmetries of the Weyl tensor, this already implies that W_p vanishes. This proves the lemma. □

To study the behavior of this operator, we define a function $w : M \to \mathbb{R}$ by

$$
w(p) = \sup\left\{ \frac{\langle \mathring{W}\eta, \eta \rangle_p}{|\eta|_p^2} \;\middle|\; \eta \in (S^2_gM)_p \right\}. \tag{4.18}
$$

Thus, $w(p)$ is the largest eigenvalue of the action $\mathring{W}\colon (S^2_gM)_p \to (S^2_gM)_p$. Lemma 4.3.2 implies that the function w is nonnegative.

The decomposition of $\mathring{R}$ allows us to estimate the smallest eigenvalue of Δ_E acting on TT-tensors in terms of the function w. From (4.2), we obtain

$$
\begin{aligned}
(\Delta_E h, h) &= \|D_1 h\|_{L^2}^2 + 2\mu\|h\|_{L^2}^2 - 4(\mathring{R}h, h) \\
&\geq 2\mu\|h\|_{L^2}^2 + 4\frac{\mu}{n-1}\|h\|_{L^2}^2 - 4(\mathring{W}h, h) \\
&\geq 2\mu\frac{n+1}{n-1}\|h\|_{L^2}^2 - 4\int_M w\cdot|h|^2 \, dV_g \\
&\geq \left[2\mu\frac{n+1}{n-1} - 4\|w\|_\infty\right]\|h\|_{L^2}^2,
\end{aligned}
$$

and similarly from (4.3),

$$\begin{aligned}(\Delta_E h, h) &= \|D_2 h\|_{L^2}^2 - \mu \|h\|_{L^2}^2 - (\mathring{R}h, h) \\ &\geq -\mu \|h\|_{L^2}^2 + \frac{\mu}{n-1} \|h\|_{L^2}^2 - (\mathring{W}h, h) \\ &\geq -\mu \frac{n-2}{n-1} \|h\|_{L^2}^2 - \int_M w \cdot |h|^2 \, dV_g \\ &\geq \left[-\mu \frac{n-2}{n-1} - \|w\|_\infty \right] \|h\|_{L^2}^2 .\end{aligned}$$

Proposition 4.3.3. *Let (M, g) be Einstein with constant μ and let λ be the smallest eigenvalue of $\Delta_E|_{TT}$. Then*

$$\lambda \geq \max \left\{ 2\mu \frac{n+1}{n-1} - 4 \|w\|_\infty , -\mu \frac{n-2}{n-1} - \|w\|_\infty \right\} .$$

As a consequence, we have

Theorem 4.3.4. *An Einstein manifold (M, g) with constant μ is stable if*

$$\|w\|_\infty \leq \max \left\{ \mu \frac{n+1}{2(n-1)} , -\mu \frac{n-2}{n-1} \right\} .$$

If the strict inequality holds, (M, g) is strictly stable.

We now give a different stability criterion which involves an integral of the function w. The main tool we use here is the Sobolev inequality which holds for Yamabe metrics.

Proposition 4.3.5 (Sobolev inequality). *Let (M, g) be a Yamabe metric in a conformal class and suppose that* $\mathrm{vol}(M, g) = 1$. *Then for any $f \in H^1(M)$,*

$$4 \frac{n-1}{n-2} \|\nabla f\|_{L^2}^2 \geq \mathrm{scal} \left\{ \|f\|_{L^p}^2 - \|f\|_{L^2}^2 \right\} \tag{4.19}$$

where $p = 2n/(n-2)$.

Proof. This follows easily from the definition of Yamabe metrics, see e.g. [IS02]. □

Remark 4.3.6. The inequality holds if f is replaced by any tensor T because of Kato's inequality

$$|\nabla |T|| \leq |\nabla T| . \tag{4.20}$$

As remarked in [LeB99, p. 329], any Einstein metric is Yamabe, so the Sobolev inequaliy holds in this case.

Theorem 4.3.7. *Let (M, g) be an Einstein manifold with positive Einstein constant μ. If*

$$\|w\|_{L^{n/2}} \leq \mu \cdot \mathrm{vol}(M, g)^{2/n} \cdot \frac{n+1}{2(n-1)} \left(\frac{4(n-1)}{n(n-2)} + 1 \right)^{-1} , \tag{4.21}$$

then (M, g) is stable. If the strict inequality holds, (M, g) is strictly stable.

Proof. Both sides of the inequality are scale-invariant, see Lemma 4.3.8 below. Therefore, we may assume $\mathrm{vol}(M,g) = 1$ from now on. First, we estimate the largest eigenvalue of the Weyl tensor action by

$$\begin{aligned}(\mathring{W}h, h)_{L^2} &\leq \int_M |w||h|^2 \, dV \\ &\leq \|w\|_{L^{n/2}} \left\||h|^2\right\|_{L^{n/(n-2)}} \\ &= \|w\|_{L^{n/2}} \|h\|^2_{L^{2n/(n-2)}} \\ &\leq \|w\|_{L^{n/2}} \left(4\frac{n-1}{\mu n(n-2)} \|\nabla h\|^2_{L^2} + \|h\|^2_{L^2}\right).\end{aligned}$$

We used the Hölder inequality and the Sobolev inequality. With the estimate obtained, we can proceed as follows:

$$\begin{aligned}(\Delta_E h, h)_{L^2} &= \|\nabla h\|^2_{L^2} - 2(\mathring{R}h, h)_{L^2} \\ &= \|\nabla h\|^2_{L^2} + 2\frac{\mu}{n-1} \|h\|^2_{L^2} - 2(\mathring{W}h, h)_{L^2} \\ &\geq \|\nabla h\|^2_{L^2} + 2\frac{\mu}{n-1} \|h\|^2_{L^2} - 2\|w\|_{L^{n/2}} \left(4\frac{n-1}{\mu n(n-2)} \|\nabla h\|^2_{L^2} + \|h\|^2_{L^2}\right) \\ &= \left(1 - 8\frac{n-1}{\mu n(n-2)} \|w\|_{L^{n/2}}\right) \|\nabla h\|^2_{L^2} + 2\left(\frac{\mu}{n-1} - \|w\|_{L^{n/2}}\right) \|h\|^2_{L^2}.\end{aligned}$$

The first term on the right hand side is nonnegative by the assumption on w. It remains to estimate $\|\nabla h\|^2_{L^2}$. This can be done by using (4.2). We have

$$\begin{aligned}\|\nabla h\|^2_{L^2} &= \|D_1 h\|^2_{L^2} + 2\mu \|h\|^2_{L^2} - 2(\mathring{R}h, h)_{L^2} \\ &\geq 2\mu \|h\|^2_{L^2} + 2\frac{\mu}{n-1} \|h\|^2_{L^2} - 2(\mathring{W}h, h)_{L^2} \\ &= 2\mu\frac{n}{n-1} \|h\|^2_{L^2} - 2(\mathring{W}h, h)_{L^2} \\ &= 2\mu\frac{n}{n-1} \|h\|^2_{L^2} - 2\|w\|_{L^{n/2}} \left(4\frac{n-1}{\mu n(n-2)} \|\nabla h\|^2_{L^2} + \|h\|^2_{L^2}\right) \\ &= 2\left(\mu\frac{n}{n-1} - \|w\|_{L^{n/2}}\right) \|h\|^2_{L^2} - 8\frac{n-1}{\mu n(n-2)} \|w\|_{L^{n/2}} \|\nabla h\|^2_{L^2},\end{aligned}$$

and therefore, $\|\nabla h\|^2_{L^2}$ can be estimated by

$$\|\nabla h\|^2_{L^2} \geq 2\left(\mu\frac{n}{n-1} - \|w\|_{L^{n/2}}\right)\left(1 + 8\frac{n-1}{\mu n(n-2)} \|w\|_{L^{n/2}}\right)^{-1} \|h\|^2_{L^2}.$$

Combining these arguments, we obtain

$$\begin{aligned}(\Delta_E h, h)_{L^2} \geq &\Bigg\{2\left(1 - 8\frac{n-1}{\mu n(n-2)} \|w\|_{L^{n/2}}\right)\left(\mu\frac{n}{n-1} - \|w\|_{L^{n/2}}\right) \cdot \\ &\left(1 + 8\frac{n-1}{\mu n(n-2)} \|w\|_{L^{n/2}}\right)^{-1} + 2\left(\frac{\mu}{n-1} - \|w\|_{L^{n/2}}\right)\Bigg\} \|h\|^2_{L^2}.\end{aligned}$$

The manifold (M,g) is stable if the right-hand side of this inequality is nonnegative. It is elementary to check that this is equivalent to

$$\|w\|_{L^{n/2}} \leq \mu\frac{n+1}{2(n-1)}\left(\frac{4(n-1)}{n(n-2)} + 1\right)^{-1}.$$

The assertion about strict stability is also immediate. □

Lemma 4.3.8. *The $L^{n/2}$-Norm of the function w is conformally invariant.*

Proof. Let $g, \tilde{g}$ be conformally equivalent, i.e. $\tilde{g} = f \cdot g$ for a smooth positive function f. Let W and $\tilde{W}$ be the Weyl tensors of the metrics g and $\tilde{g}$, respectively. We have $\mathrm{tr}_{\tilde{g}} h = f^{-1}\mathrm{tr}_g h$ for any symmetric $(0,2)$-tensor. Thus,

$$S^2_{\tilde{g}}M = S^2_g M,$$

so the operators $\mathring{\tilde{W}}$ and $\mathring{W}$ are acting on the same space of tensors. It is well-known that $\tilde{W} = f \cdot W$ when considered as $(0,4)$-tensors. Therefore,

$$\langle \mathring{\tilde{W}} h, h\rangle_{\tilde{g}} = f^{-3}\langle \mathring{W} h, h\rangle_g.$$

Furthermore, we have

$$|h|^2_{\tilde{g}} = f^{-2}|h|^2_g, \qquad dV_{\tilde{g}} = f^{n/2}\, dV_g.$$

We now see that the largest eigenvalue of the Weyl-tensor action transforms as $\tilde{w} = f^{-1}w$ and

$$\|\tilde{w}\|^{2/n}_{L^{n/2}(\tilde{g})} = \int_M \tilde{w}^{n/2}\, dV_{\tilde{g}} = \int_M w^{n/2}\, dV_g = \|w\|^{2/n}_{L^{n/2}(g)},$$

which shows the lemma. □

Corollary 4.3.9. *Let (M,g) be a Riemannian manifold and let $Y([g])$ be the Yamabe constant of the conformal class of g. If*

$$\|w\|_{L^{n/2}(g)} \leq Y([g])\frac{n+1}{2n(n-1)} \cdot \left(\frac{4(n-1)}{n(n-2)} + 1\right)^{-1}, \tag{4.22}$$

any Einstein metric in the conformal class of g is stable.

Proof. Suppose that $\tilde{g} \in [g]$ is Einstein. By Lemma 4.3.8,

$$\|\tilde{w}\|_{L^{n/2}(\tilde{g})} = \|w\|_{L^{n/2}(g)}.$$

We know that $\tilde{g}$ is a Yamabe metric in the conformal class of g. By the definition of the Yamabe constant, the Einstein constant of $\tilde{g}$ equals

$$\mu = \frac{1}{n} \cdot Y([g]) \cdot \mathrm{vol}(M,g)^{2/n},$$

which yields

$$\|\tilde{w}\|_{L^{n/2}(\tilde{g})} \leq \mu \cdot \mathrm{vol}(M,g)^{2/n} \cdot \frac{n+1}{2(n-1)} \cdot \left(\frac{4(n-1)}{n(n-2)} + 1\right)^{-1}.$$

The assertion now follows from Theorem 4.3.7. □

Now we give upper estimates for the values of the function w:

Lemma 4.3.10. *Let (M, g) be Einstein and $p \in M$. Let $W_{min} = \min W_{ijji}$ and $W_{max} = \max W_{ijji}$ where the minimum (resp. maximum) is taken over all orthonormal bases of T_pM. Then*

$$w(p) \leq \min\{(n-2)W_{max}, -nW_{min}\}. \tag{4.23}$$

Proof. For the sake of completeness, we give the proof although it is completely analogous to the proof of Lemma 4.1.2. Let κ be an eigenvalue of $\mathring{W}$ and choose $\eta \in (S^2_g M)_p$ such that $\mathring{W}\eta = \kappa\eta$. Choose an orthonormal basis $\{e_1, \ldots, e_n\}$ of eigenvectors of η with eigenvalues $\lambda_1, \ldots, \lambda_n$ such that $\lambda_1 = \sup|\lambda_i|$ and $\sum \lambda_i = 0$. Then

$$\kappa\lambda_1 = \kappa\eta(e_1, e_1) = (\mathring{W}\eta)(e_1, e_1) = \sum_{i,j} W(e_i, e_1, e_1, e_j)h(e_j, e_i) = \sum_i W_{i11i}\lambda_i.$$

Thus,

$$\begin{aligned}\kappa\lambda_1 &= \sum_{i\neq 1} W_{max}\lambda_i - \sum_{i\neq 1}(W_{max} - W_{i11i})\lambda_i \\ &\leq -\lambda_1 W_{max} + \lambda_1 \sum_{i\neq 1}(W_{max} - W_{i11i}) \\ &= (n-2)W_{max}\lambda_1,\end{aligned}$$

where we used the fact that W is trace-free. Furthermore,

$$\begin{aligned}\kappa\lambda_1 &= \sum_{i\neq 1} W_{min}\lambda_i + \sum_{i\neq 1}(W_{i11i} - W_{min})\lambda_i \\ &\leq -\lambda_1 W_{min} + \lambda_1 \sum_{i\neq 1}(W_{i11i} - W_{min}) \\ &= -nW_{min}\lambda_1,\end{aligned}$$

which finishes the proof. □

Observe that $W_{min} = K_{min} - \frac{\mu}{n-1}$ and that $W_{max} = K_{max} - \frac{\mu}{n-1}$ for Einstein manifolds with constant μ. By using the Cauchy-Schwarz inequality, we have

Lemma 4.3.11. *Let (M, g) be a Riemannian manifold. Then*

$$w(p) \leq |W|_p. \tag{4.24}$$

There are also attempts to prove stability criterions involving an eigenvalue estimate of the Weyl curvature operator. Recall that by its symmetries, the Weyl tensor can be considered as a self-adjoint operator acting on 2-forms by defining

$$\langle \hat{W}(X \wedge Y), Z \wedge W\rangle = W(Y, X, Z, W).$$

We call $\hat{W}\colon \Gamma(\Lambda^2 M) \to \Gamma(\Lambda^2 M)$ the Weyl curvature operator. Let $\overline{w}(p)$ be its largest eigenvalue at $p \in M$.

Theorem 4.3.12 ([IN05]). *Let (M, g) be a compact, connected oriented Einstein manifold with negative Einstein constant μ. If*

$$\sup_{p \in M} \overline{w}(p) < -\frac{\mu}{n-1}, \tag{4.25}$$

then (M, g) is strictly stable.

However, the proof uses the very rough estimate

$$W_{max} \leq \overline{w}(p). \tag{4.26}$$

In fact, the Theorem follows directly from combining Theorem 4.3.4, Lemma 4.3.10 and (4.26). Therefore, it seems not convenient to formulate stability criterions in terms of the Weyl curvature operator because we find no direct way to estimate $w(p)$ in terms of $\overline{w}(p)$ without using (4.26).

4.4 Isolation Results of the Weyl Curvature Tensor

In the last section, we have shown that an Einstein manifold is stable if its Weyl tensor is small enogh in a certain sense. The smallness of the tensor was expressed in serveral ways. However, we have to be careful. There exists various results (see [Mut69; Sin92; GL99; IS02]) which state that if the Weyl tensor of an Einstein metric is small enough, it vanishes identically. A strong result of this form is the following

Theorem 4.4.1 ([IS02]). *Let (M, g) be a compact connected, oriented Einstein-manifold, $n \geq 4$, with positive Einstein constant μ and of unit-volume. Then there exists a constant $C(n)$, depending only on n, such that if the inequality $\|W\|_{L^{n/2}} < C(n)n\mu$ holds, then $W = 0$ so that (M, g) is a finite isometric quotient of the sphere.*

More precisely, they put $C(n)$ as

$$C(n) = \begin{cases} \frac{n-2}{4(n-1)} C_n & \text{if } 4 \leq n \leq 9, \\ \frac{2}{n} C_n & \text{if } n \geq 10. \end{cases}$$

The constant C_n appears in the estimate

$$|\langle \{W, W\}, W \rangle| \leq C_n^{-1} |W|^3,$$

and $\{W, W\}$ is a $(0, 4)$-tensor quadratic in W which appears in the formula

$$0 = \nabla^* \nabla W + 2\mu W + \{W, W\}.$$

We are interested in the value of C_n. Therefore, we compute the coordinate expression for $\{W, W\}$. On any Einstein manifold, we have $\nabla W = \nabla R$ since the difference $R - W$ is a parallel tensor. Moreover

$$\nabla_i W_{jklm} + \nabla_j W_{kilm} + \nabla_k W_{ijlm} = 0,$$
$$\nabla_i W_{ijkl} = 0,$$

where the components of W are taken with respect to an orthonormal frame. The first equation is just the second Bianchi identity and the second equation follows from contracting the first. Using these two properties, we obtain

$$\sum_i \{-\nabla^2_{ii} W_{jklm} + (\nabla^2_{ij} - \nabla^2_{ji}) W_{iklm} - (\nabla^2_{ik} - \nabla^2_{ki}) W_{ijlm}\} = 0.$$

By the Ricci identity and the first Bianchi identity,

$$\begin{aligned}
\sum_i (\nabla^2_{ij} - \nabla^2_{ji}) W_{iklm} \\
&= \sum_{i,n} \{R_{jiin} W_{nklm} + R_{jikn} W_{inlm} + R_{jiln} W_{iknm} + R_{jimn} W_{ikln}\} \\
&= \mu W_{jklm} + \sum_{i,n} \{W_{jikn} W_{inlm} + W_{jiln} W_{iknm} + W_{jimn} W_{ikln}\} \\
&\quad + \frac{\mu}{n-1} (W_{kjlm} + W_{lkjm} + W_{mklj}) \\
&= \mu W_{jklm} + \sum_{i,n} \{W_{jikn} W_{inlm} + W_{jiln} W_{iknm} + W_{jimn} W_{ikln}\}.
\end{aligned}$$

An analogous formula is valid for $(\nabla^2_{ik} - \nabla^2_{ki}) W_{ijlm}$. Therefore,

$$0 = \nabla^* \nabla W + 2\mu W + \{W, W\},$$

and $\{W, W\}$ consists of six summands of the form $\sum_{i,n} W_{jikn} W_{inlm}$. It is immediate that

$$|\langle \{W, W\}, W \rangle| \leq 6|W|^3, \tag{4.27}$$

so $C_n = 1/6$ is an appropriate choice since (4.27) seems to be not far away from the optimum. This yields

$$\|W\|_{L^{n/2}} \geq \begin{cases} \frac{n(n-2)}{24(n-1)} \mu & \text{if } 4 \leq n \leq 9, \\ \frac{1}{3} \mu & \text{if } n \geq 10. \end{cases} \tag{4.28}$$

Recall Theorem 4.3.7. There, we proved stability for a small $L^{n/2}$-norm of the function w. From Lemma 4.3.11, we conclude that a positive Einstein manifold of unit volume is stable, if

$$\|W\|_{L^{n/2}} \leq \mu \cdot \frac{n+1}{2(n-1)} \left(\frac{4(n-1)}{n(n-2)} + 1 \right)^{-1}. \tag{4.29}$$

A comparison of the last two inequalities shows that there exists a small gap where the inequality (4.29) works.

By different techniques, Gursky and LeBrun proved a gap theorem for the Weyl tensor, which only holds in dimension 4. For any oriented Riemannian 4-manifold the Weyl-tensor orthogonally splits as $W = W^+ + W^-$ where W^+, W^- is the self-dual (resp. anti-self-dual) part of the Weyl tensor.

Theorem 4.4.2 ([GL99]). *Let (M, g) be a compact oriented Einstein 4-manifold with* $\mathrm{scal} > 0$ *and $W^+ \not\equiv 0$. Then*

$$\int_M |W^+|^2 \, dV \geq \int_M \frac{\mathrm{scal}^2}{6} \, dV,$$

with equality if and only if $\nabla W^+ \equiv 0$.

Here, we translated this result to our norm convention for curvature tensors which differs from the one in [GL99] by a factor 1/4. By changing orientation, the roles of W^+ and W^- interchange and we see that the same gap theorem also holds for W^-. Therefore, if $W^\pm \neq 0$,

$$\|W^\pm\|_{L^2}^2 \geq \int_M \frac{\mathrm{scal}^2}{6}\, dV = \mathrm{vol}(M,g)\frac{8}{3}\mu^2.$$

If $W \not\equiv 0$, either W^+ or W^- is not vanishing, so the same gap holds for $\|W\|_{L^2}$. By passing to the orientation covering, we see that the same gap also holds for non-orientable Einstein manifolds. This gap is much larger than the one proven by Itoh and Satoh. If (M,g) is of unit-volume, we have $\|W\|_{L^2} \geq \sqrt{\frac{8}{3}}\mu$ while (4.28) yields $\|W\|_{L^2} \geq \frac{\mu}{9}$. In fact, it shows that stability criterion (4.29) is useless in dimension 4, since this requires the Weyl tensor to satisfy $\|W\|_{L^2} \leq \frac{\mu}{3}$.

4.5 Six-dimensional Einstein Manifolds

In this section, we compute an explicit representation of the Gauss-Bonnet formula for six-dimensional Einstein manifolds. We use this representation to show a stability criterion for Einstein manifolds involving the Euler characteristic. The generalized Gauss-Bonnet formula for a compact Riemannian manifold (M,g) of dimension $n = 2m$ is

$$\chi(M) = \frac{(-1)^m}{2^{3m}\pi^m m!}\int_M \Psi_g\, dV.$$

The function Ψ_g is defined as

$$\Psi_g = \sum_{\sigma,\tau\in S_m} \mathrm{sgn}(\sigma)\mathrm{sgn}(\tau) R_{\sigma(1)\sigma(2)\tau(1)\tau(2)} \cdots R_{\sigma(n-1)\sigma(n)\tau(n-1)\tau(n)},$$

where the coefficients are taken with respect to an orthonormal basis (see e.g. [Zhu00]). In dimension four, this yields the nice formula

$$\chi(M) = \frac{1}{32\pi^2}\int_M (|W|^2 + |Sc|^2 - |U|^2)\, dV \tag{4.30}$$

(see also [Bes08, p. 161]). Here, $Sc = \frac{\mathrm{scal}}{2n(n-1)} g \owedge g$ is the scalar part and $U = \frac{1}{n-2}\mathrm{Ric}^0 \owedge g$ is the traceless Ricci part of the curvature tensor. Due to different conventions for the norm of curvature tensors, formula (4.30) often appers with the factor $\frac{1}{8\pi^2}$ instead of $\frac{1}{32\pi^2}$. On Einstein manifolds, we have $U = 0$ and the Gauss-Bonnet formula simplifies to

$$\chi(M) = \frac{1}{32\pi^2}\int_M \left(|W|^2 + \frac{8}{3}\mu^2\right) dV \tag{4.31}$$

where μ is the Einstein constant. As a nice consequence, we obtain a topological condition for the existence of Einstein metrics which is due to Berger.

Theorem 4.5.1 ([Ber65]). *Every compact 4-manifold carrying an Einstein metric g satisfies the inequality*

$$\chi(M) \geq 0.$$

Moreover, $\chi(M) = 0$ *if and only if* (M,g) *is flat.*

Another consequence of (4.31) is the following: Let (M,g) be of unit volume. Then there exists a constant $C>0$ such that, if $\mu \geq C \cdot \sqrt{\chi(M)}$, the Weyl curvature satisfies $\|W\|_{L^2} \leq \frac{1}{3}\mu$. This implies stability by Theorem 4.3.7 and Lemma 4.3.11. Unfortunately, the same condition on the Weyl tensor already implies that it vanishes, as we discussed in the last section. Thus, the assertion is of no use here.

In dimension six, an explicit representation of the Gauss-Bonnet formula is given by

$$\begin{aligned}\chi(M) =&\frac{1}{384\pi^3}\int_M \{\mathrm{scal}^3 - 12\mathrm{scal}|\mathrm{Ric}|^2 + 3\mathrm{scal}|R|^2 + 16\langle \mathrm{Ric}, \mathrm{Ric}\circ\mathrm{Ric}\rangle \\ &- 24\mathrm{Ric}^{ij}\mathrm{Ric}^{kl}R_{ikjl} - 24\mathrm{Ric}_i^{\ j}R^{iklm}R_{jklm} + 8R^{ijkl}R_{imkn}R_{j\ l}^{\ n\ m} \\ &- 2R^{ijkl}R_{ij}^{\ \ mn}R_{klmn}\}\ dV\end{aligned}$$

(see [Sak71, Lemma 5.5]). When (M,g) is Einstein, this integral is equal to

$$\begin{aligned}\chi(M) =&\frac{1}{384\pi^3}\int_M \{24\mu^3 - 6\mu|R|^2 + 8R^{ijkl}R_{imkn}R_{j\ l}^{\ n\ m} \\ &- 2R^{ijkl}R_{ij}^{\ \ mn}R_{klmn}\}\ dV.\end{aligned} \tag{4.32}$$

Lemma 4.5.2. *If (M,g) is a compact Einstein manifold with constant μ,*

$$\|\nabla R\|_{L^2}^2 = -\int_M \{4R^{ijkl}R_{i\ k}^{\ m\ n}R_{jnlm} + 2R^{ijkl}R_{ij}^{\ \ mn}R_{klmn} + 2\mu|R|^2\}\ dV.$$

Proof. This is [Sak71, (2.15)] in the special case of Einstein metrics. □

Note that we translated the formulas from [Sak71] to our sign convention for the curvature tensor.

Proposition 4.5.3. *Let (M,g) be an Einstein six-manifold with constant μ. Then*

$$\chi(M) = \frac{1}{384\pi^3}\int_M \left\{-\frac{14}{5}\mu|W|^2 - 2|\nabla W|^2 + \frac{144}{25}\mu^3 + 48\mathrm{tr}(\hat{W}^3)\right\}\ dV.$$

Here, $\hat{W}^3 = \hat{W}\circ\hat{W}\circ\hat{W}$, where $\hat{W}$ is the Weyl curvature operator acting on 2-forms.

Proof. By Lemma 4.5.2, (4.32) can be rewritten as

$$384\pi^3\chi(M) = \int_M \{24\mu^3 - 10\mu|R|^2 - 2|\nabla R|^2 - 6R^{ijkl}R_{ij}^{\ \ mn}R_{klmn}\}\ dV.$$

Moreover, $\nabla W = \nabla R$ because the difference $R - W = Sc$ is a parallel tensor. Thus,

$$\begin{aligned}384\pi^3\chi(M) &= \int_M \{24\mu^3 - 10\mu|R|^2 - 2|\nabla W|^2 - 6R^{ijkl}R_{ij}^{\ \ mn}R_{klmn}\ dV\} \\ &= \int_M \{24\mu^3 - 10\mu(|Sc|^2 + |W|^2) - 2|\nabla W|^2 - 6R^{ijkl}R_{ij}^{\ \ mn}R_{klmn}\}\ dV \\ &= \int_M \left\{24\mu^3 - 10\mu\left(\frac{12\mu^2}{5} + |W|^2\right) - 2|\nabla W|^2 - 6R^{ijkl}R_{ij}^{\ \ mn}R_{klmn}\right\}\ dV \\ &= \int_M \{-10\mu|W|^2 - 2|\nabla W|^2 - 6R^{ijkl}R_{ij}^{\ \ mn}R_{klmn}\}\ dV.\end{aligned}$$

Now we analyse the last term on the right hand side. Recall that the Riemann curvature operator $\hat{R}$ and the Weyl curvature operator $\hat{W}$ are defined by

$$\langle \hat{R}(X \wedge Y), Z \wedge V \rangle = R(Y, X, Z, V),$$
$$\langle \hat{W}(X \wedge Y), Z \wedge V \rangle = W(Y, X, Z, V).$$

Let $\{e_1, \ldots, e_n\}$ be a local orthonormal frame of TM. Then $\{e_i \wedge e_j\}$, $i < j$ is a local orthonormal frame of $\Lambda^2 M$. A straightforward calculation shows

$$\begin{aligned} -6 \sum_{i,j,k,l,m,n} R_{ijkl} R_{ijmn} R_{klmn} &= -48 \sum_{i<j,k<l,m<n} R_{ijkl} R_{ijmn} R_{klmn} \\ &= 48 \sum_{i<j} g(\hat{R}(\hat{R}(\hat{R}(e_i \wedge e_j))), e_i \wedge e_j) = 48 \mathrm{tr} \hat{R}^3, \end{aligned}$$

where the coefficients of R are taken with respect to the orthonormal frame. The decomposition (4.17) of the $(0,4)$-curvature tensor induces the decomposition $\hat{R} = \hat{W} + \frac{\mu}{5} \mathrm{id}_{\Lambda^2 M}$. This yields

$$\begin{aligned} 48 \mathrm{tr} \hat{R}^3 &= 48 \left\{ \mathrm{tr}(\hat{W}^3) + 3 \frac{\mu}{5} \mathrm{tr}(\hat{W}^2) + 3 \frac{\mu^2}{25} \mathrm{tr} \hat{W} + \frac{\mu^3}{125} \mathrm{tr}(\mathrm{id}_{\Lambda^2 M}) \right\} \\ &= 48 \mathrm{tr}(\hat{W}^3) + \frac{36}{5} \mu |W|^2 + \frac{144}{25} \mu^3. \end{aligned}$$

Therefore, we have

$$\begin{aligned} 384 \pi^3 \chi(M) &= \int_M \{-10\mu |W|^2 - 2|\nabla W|^2 + \frac{144}{25} \mu^3 + \frac{36}{5} \mu |W|^2 + 48 \mathrm{tr}(\hat{W}^3)\} \, dV \\ &= \int_M \{-\frac{14}{5} \mu |W|^2 - 2|\nabla W|^2 + \frac{144}{25} \mu^3 + 48 \mathrm{tr}(\hat{W}^3)\} \, dV, \end{aligned}$$

which finishes the proof. □

Theorem 4.5.4. *Let (M, g) be a positive Einstein six-manifold with constant μ and* $\mathrm{vol}(M) = 1$. *If*

$$\frac{1}{25} \left(144 - \frac{12 \cdot 7^2 \cdot 3^2}{5 \cdot 11^2} \right) \mu^3 \leq 384 \pi^3 \chi(M) - 48 \int_M \mathrm{tr}(\hat{W}^3) \, dV,$$

then (M, g) is strictly stable.

Proof. By the Sobolev inequality,

$$\|W\|_{L^3}^2 \leq \frac{5}{6\mu} \|\nabla W\|_{L^2}^2 + \|W\|_{L^2}^2 .$$

Therefore we have, by Proposition 4.5.3

$$\begin{aligned} 384 \pi^3 \chi(M) &= -\frac{14}{5} \mu \|W\|_{L^2}^2 - 2 \|\nabla W\|_{L^2}^2 + \frac{144}{25} \mu^3 + 48 \int_M \mathrm{tr}(\hat{W}^3) \, dV \\ &< -\frac{12}{5} \mu \|W\|_{L^2}^2 - 2 \|\nabla W\|_{L^2}^2 + \frac{144}{25} \mu^3 + 48 \int_M \mathrm{tr}(\hat{W}^3) \, dV \\ &\leq -\frac{12}{5} \mu \|W\|_{L^3}^2 + \frac{144}{25} \mu^3 + 48 \int_M \mathrm{tr}(\hat{W}^3) \, dV. \end{aligned}$$

Now if μ satisfies the estimate of the statement in the theorem, we obtain

$$\begin{aligned}\frac{12}{5}\mu \|W\|_{L^3}^2 &< \frac{144}{25}\mu^3 - 384\pi^3\chi(M) + 48\int_M \operatorname{tr}(\hat{W}^3)\, dV \\ &\le \frac{144}{25}\mu^3 - \frac{1}{25}\left(144 - \frac{12\cdot 7^2\cdot 3^2}{5\cdot 11^2}\right)\mu^3 = \frac{12\mu}{5}\frac{7^2\cdot 3^2}{5^2\cdot 11^2}\mu^2,\end{aligned}$$

which is equivalent to

$$\|W\|_{L^3} < \frac{7\cdot 3}{5\cdot 11}\mu.$$

By Theorem 4.3.7 and Lemma 4.3.11, (M, g) is strictly stable. □

Remark 4.5.5. Mind the fact that this stability criterion is not ruled out by isolation results.

4.6 Kähler Manifolds

Here, we prove stability criterions for Kähler-Einstein manifolds in terms of the Bochner curvature tensor, which is an analogue of the Weyl tensor.

Definition 4.6.1. Let (M, g) be a Riemannian manifold of even dimension. An almost complex structure on M is an endorphism $J : TM \to TM$ such that $J^2 = -\mathrm{id}_{TM}$. If J is parallel and g is hermitian, i.e. $g(JX, JY) = g(X, Y)$, we call the triple (M, g, J) a Kähler manifold. If (M, g) is Einstein, we call (M, g, J) Einstein-Kähler.

If J is parallel, $R(X,Y)JZ = JR(X,Y)Z$ and we get an additional symmetry for the $(0,4)$-cuvature tensor, namely

$$R(JX, JY, Z, W) = R(X, Y, JZ, JW) = R(X, Y, Z, W).$$

We say that R is hermitian. The bundle of traceless symmetric $(0,2)$-tensors splits into hermitian and skew-hermitian ones, i.e. $S^2_g M = H_1 \oplus H_2$, where

$$\begin{aligned}H_1 &= \left\{h \in S^2_g M \mid h(X,Y) = h(JX, JY)\right\}, \\ H_2 &= \left\{h \in S^2_g M \mid h(X,Y) = -h(JX, KY)\right\}.\end{aligned}$$

Stability of Kähler-Einstein manifolds was studied in [Koi83; IN05; DWW07]. We sketch the ideas of [Koi83] in the following. It turns out that the Einstein operator preserves the bundle splitting $\Gamma(H_1) \oplus \Gamma(H_2)$. Therefore to show that a Kähler-Einstein manifold is stable it is sufficient to show that the restriction of Δ_E to the subspaces $\Gamma(H_1)$ and $\Gamma(H_2)$ is positive semidefinite, respectively. In fact, we can use the Kähler structure to conjugate the Einstein operator to other operators. If $h_1 \in H_1$, we define a 2-form by

$$\phi(X, Y) = h_1 \circ J(X, Y) = h_1(X, J(Y)).$$

We have

$$\Delta_H \phi = (\Delta_E h_1) \circ J + 2\mu\phi, \tag{4.33}$$

where Δ_H is the Hodge Laplacian on 2-forms and μ is the Einstein constant. Since Δ_H is nonnegative, Δ_E is nonnegative on $\Gamma(H_1)$, if $\mu \leq 0$. For $h_2 \in H_2$, we define a symmetric endomorphism $I : TM \to TM$ by

$$g \circ I = h_2 \circ J,$$

and since $IJ + JI = 0$, we may consider I as a $T^{1,0}M$-valued 1-form of type $(0,1)$. We have the formula

$$g \circ (\Delta_C I) = (\Delta_E h_2) \circ J, \tag{4.34}$$

where Δ_C is the complex Laplacian. Thus, the restriction of the Einstein operator to $\Gamma(H_2)$ is always nonnegative, since Δ_C is. As a consequence, we have

Corollary 4.6.2 ([DWW07]). *Any compact Kähler-Einstein manifold with nonpositive Einstein constant is stable.*

Remark 4.6.3. This is not true for positive Kähler-Einstein manifolds. The product of two positive Kähler-Einstein manifolds is unstable.

Using (4.33) and (4.34), $\dim(\ker \Delta_E|_{TT})$ can be expressed in terms of certain cohomology classes (see [Koi83] or [Bes08, Proposition 12.98]). Moreover, integrability of infinitesimal Einstein deformations can be related to integrability of infinitesimal complex deformations (see [Koi83; IN05]).

We discuss conditions under which a Kähler-Einstein manifold is strictly stable in the nonpositive case and stable in the positive case. This can be described in terms of the Bochner curvature tensor which has similar properties as the Weyl tensor.

Definition 4.6.4 (Bochner curvature tensor). Let (M, g, J) be Kähler and let $\omega(X,Y) = g(J(X),Y)$ be the Kähler form. The Bochner curvature tensor is defined by

$$\begin{aligned} B =& R + \frac{\text{scal}}{2(n+2)(n+4)} \{g \owedge g + \omega \owedge \omega - 4\omega \otimes \omega\} \\ &- \frac{1}{n+4} \{\text{Ric} \owedge g + (\text{Ric} \circ J) \owedge \omega - 2(\text{Ric} \circ J) \otimes \omega - 2\omega \otimes (\text{Ric} \circ J)\} \end{aligned}$$

(see e.g. [IK04, p. 229]).

The Bochner curvature tensor posesses the same symmetries as the Riemann tensor and in addition,

$$B(JX, JY, Z, W) = B(X, Y, JZ, JW) = B(X, Y, Z, W),$$

$$\sum_{i=1}^{n} B(X, e_i, e_i, Y) = 0.$$

If (M, g) is Kähler-Einstein, the Bochner tensor is

$$B = R - \frac{\mu}{2(n+2)} \{g \owedge g + \omega \owedge \omega - 4\omega \otimes \omega\},$$

where μ is the Einstein constant (see e.g. [IK04; IN05] and mind the different sign convention for the curvature tensor). This implies the relation

$$B(X, J(X), J(X), X) = R(X, J(X), J(X), X) - \frac{4\mu}{n+2}|X|^4.$$

In particular, if the Bochner tensor vanishes, the holomorphic setional curvature, i.e. the sectional curvature of all planes spanned by X and $J(X)$ is constant. The Bochner tensor acts naturally on symmetric $(0,2)$-tensors by

$$\mathring{B}h(X,Y) = \sum_{i,j=n}^{n} B(e_i,X,Y,e_j)h(e_i,e_j),$$

where $\{e_1,\dots,e_n\}$ is an orthonormal basis. Let

$$b^+(p) = \left\{ \frac{\langle \mathring{B}\eta,\eta\rangle}{|\eta|^2} \,\middle|\, \eta \in (H_1)_p \right\}.$$

For Kähler-Einstein manifolds with negative Einstein constant, it was proven by M. Itoh and T. Nakagawa that they are strictly stable if the Bochner tensor is small.

Theorem 4.6.5 ([IN05]). *Let (M,g,J) be a compact Kähler-Einstein manifold with negative Einstein constant μ. If the Bochner curvature tensor satisfies*

$$\sup_p b^+(p) < -\mu\frac{n}{n+2}, \tag{4.35}$$

then g is strictly stable.

However, an error occured in the calculations and the result is slightly different. Therefore, lets redo the proof. By straightforward calculation,

$$\langle \mathring{R}h,h\rangle = \langle \mathring{B}h,h\rangle - \frac{\mu}{n+2}\{|h|^2 - 3\sum_{i,j} h(e_i,e_j)h(J(e_i),J(e_j))\}. \tag{4.36}$$

In particular,

$$\langle \mathring{R}h_1,h_1\rangle = \langle \mathring{B}h_1,h_1\rangle + 2\frac{\mu}{n+2}|h_1|^2$$

for $h_1 \in H_1$ and

$$\langle \mathring{R}h_2,h_2\rangle = \langle \mathring{B}h_2,h_2\rangle - 4\frac{\mu}{n+2}|h_2|^2$$

for $h_2 \in H_2$. By (4.33), Δ_E is positive definite on $\Gamma(H_1)$ so it remains to consider the part in $\Gamma(H_2)$. By (4.3),

$$\begin{aligned}(\Delta_E h_2,h_2)_{L^2} &= \|D_2h_2\|_{L^2}^2 - \mu\|h_2\|_{L^2}^2 - (h_2,\mathring{R}h_2)_{L^2} + \|\delta h_2\|_{L^2}^2 . \\ &\geq -\mu\|h_2\|_{L^2}^2 - (h_2,\mathring{R}h_2)_{L^2} \\ &\geq -\mu\|h_2\|_{L^2}^2 - (h_2,\mathring{B}h_2)_{L^2} + \frac{4\mu}{n+2}\|h_2\|_{L^2}^2 \\ &\geq -\mu\frac{n-2}{n+2}\|h_2\|_{L^2}^2 - \sup_p b^+(p)\|h_2\|_{L^2}^2 .\end{aligned}$$

Remark 4.6.6. Theorem 4.6.5 is true if we replace (4.35) by

$$\sup_p b^+(p) < -\mu\frac{n-2}{n+2}. \tag{4.37}$$

Now, let us turn to positive Kähler-Einstein manifolds. We will use Bochner formula (4.2). Unfortunately, we cannot make use of the vector bundle splitting $S^2_gM = H_1 \oplus H_2$. In order to apply (4.2), we need the condition $\delta h = 0$, which is not preserved by the splitting into hermitian and skew-hermitian tensors. Let

$$b(p) = \sup \left\{ \frac{\langle \mathring{B}\eta, \eta \rangle}{|\eta|^2} \;\middle|\; \eta \in (S^2_gM)_p \right\}. \tag{4.38}$$

Since the trace of the Bochner tensor vanishes, $\mathring{B}\colon (S^2_gM)_p \to (S^2_gM)_p$ has also vanishing trace (this follows from the same arguments as used in the proof of Lemma 4.3.2). Thus, b is nonnegative.

Theorem 4.6.7. *Let (M, g, J) be Kähler-Einstein with positive Einstein constant μ. If*

$$\|b\|_{L^\infty} \leq \frac{\mu(n-2)}{2(n+2)},$$

then (M, g) is stable.

Proof. Let $h \in TT$. By (4.36) and the Cauchy-Schwarz inequality,

$$\langle \mathring{R}h, h \rangle \leq \langle \mathring{B}h, h \rangle + 2\frac{\mu}{n+2}|h|^2.$$

Using (4.2), we therefore obtain

$$\begin{aligned}(\Delta_E h, h)_{L^2} &= \|D_1 h\|^2_{L^2} + 2\mu \|h\|^2_{L^2} - 4(h, \mathring{R}h)_{L^2} \\ &\geq 2\mu \|h\|^2_{L^2} - 4(h, \mathring{B}h)_{L^2} - 8\frac{\mu}{n+2} \|h\|^2_{L^2} \\ &\geq 2\mu\frac{n-2}{n+2} \|h\|^2_{L^2} - 4 \|b\|_{L^\infty} \|h\|^2_{L^2}.\end{aligned}$$

Under the assumptions of the theorem, $\Delta_E|_{TT}$ is nonnegative. □

We also prove a stability criterion involving the $L^{n/2}$-norm of b:

Theorem 4.6.8. *Let (M, g, J) be a positive Kähler-Einstein manifold with constant μ and* $\mathrm{vol}(M, g) = 1$. *If the function b satisfies*

$$\|b\|_{L^{n/2}} \leq \mu \cdot \frac{(n-2)}{2(n+2)} \left(\frac{4(n-1)}{n(n+2)} + 1 \right)^{-1},$$

then (M, g) is stable.

Proof. The proof is very similar to that of Theorem 4.3.7. Let $h \in TT$. By assumtion, (M, g) is a Yamabe metric. Thus, we can use the Sobolev inequality and we get

$$\begin{aligned}(\mathring{B}h, h)_{L^2} = \int_M \langle \mathring{B}h, h \rangle &\leq \int_M b|h|^2 \, dV \\ &\leq \|b\|_{L^{n/2}} \|h\|^2_{L^{2n/n-2}} \\ &\leq \|b\|_{L^{n/2}} \left(\frac{4(n-1)}{\mu n(n-2)} \|\nabla h\|^2_{L^2} + \|h\|^2_{L^2} \right).\end{aligned}$$

By the above,

$$\begin{aligned}(\Delta_E h, h)_{L^2} &= \|\nabla h\|_{L^2}^2 - 2(\mathring{R}h, h)_{L^2} \\ &\geq \|\nabla h\|_{L^2}^2 - 2(\mathring{B}h, h)_{L^2} - \frac{4\mu}{n+2}\|h\|_{L^2}^2 \\ &\geq \|\nabla h\|_{L^2}^2 - 2\|b\|_{L^{n/2}}\left(\frac{4(n-1)}{\mu n(n-2)}\|\nabla h\|_{L^2}^2 + \|h\|_{L^2}^2\right) - \frac{4\mu}{n+2}\|h\|_{L^2}^2 \\ &= \left(1 - \frac{8(n-1)}{\mu n(n-2)}\|b\|_{L^{n/2}}\right)\|\nabla h\|_{L^2}^2 - 2\|b\|_{L^{n/2}}\|h\|_{L^2}^2 - \frac{4\mu}{n+2}\|h\|_{L^2}^2 .\end{aligned}$$

The first term on the right hand side is nonnegative by the assumption on b. To estimate $\|\nabla h\|_{L^2}^2$, we rewrite (4.2) to get

$$\begin{aligned}\|\nabla h\|_{L^2}^2 &= \|D_1 h\|_{L^2}^2 + 2\mu\|h\|_{L^2}^2 - 2(h, \mathring{R}h)_{L^2} \\ &\geq 2\mu\frac{n}{n+2}\|h\|_{L^2}^2 - 2(h, \mathring{B}h)_{L^2} \\ &\geq 2\mu\frac{n}{n+2}\|h\|_{L^2}^2 - 2\|b\|_{L^{n/2}}\left(\frac{4(n-1)}{\mu n(n-2)}\|\nabla h\|_{L^2}^2 + \|h\|_{L^2}^2\right).\end{aligned}$$

Thus,

$$\|\nabla h\|_{L^2}^2 \geq 2\left(\mu\frac{n}{n+2} - \|b\|_{L^{n/2}}\right)\left(1 + \frac{8(n-1)}{\mu n(n-2)}\right)^{-1}\|h\|_{L^2}^2 .$$

By combining these arguments,

$$\begin{aligned}(\Delta_E h, h)_{L^2} \geq \Bigg\{ & 2\left(\mu\frac{n}{n+2} - \|b\|_{L^{n/2}}\right)\left(1 - \frac{8(n-1)}{\mu n(n-2)}\|b\|_{L^{n/2}}\right) \\ & \left(1 + \frac{8(n-1)}{\mu n(n-2)}\right)^{-1} - 2\|b\|_{L^{n/2}} - \frac{4\mu}{n+2}\Bigg\}\|h\|_{L^2}^2 ,\end{aligned}$$

and the right-hand side is nonnegative if the above assumption holds. □

By the Cauchy-Schwarz inequality, we clearly have

$$b(p) \leq |B|_p. \tag{4.39}$$

Remark 4.6.9. As for the Weyl tensor, there also exist isolation results for the $L^{n/2}$-norm of the Bochner tensor, see [IK04, Theorem A]. The methods are similar to those of [IS02] and for the constant C_n appearing in formula (24) of [IK04], the value 1/6 (as in Section 4.4) seems to be not too far away from the optimum. A criterion combining Theorem 4.6.8 and (4.39) is not ruled out by these results, if $n \geq 5$. If $n = 4$, $B = W^-$ (see [IK04, p. 232]). Then Theorem 4.4.2 applies and this criterion is ruled out.

Chapter 5

Ricci Flow and negative Einstein Metrics

5.1 Introduction

The Ricci flow was first introduced by Hamilton in [Ham82]. In this section, we summarize some facts about Ricci flow which are relevant to the rest of the chapter. More details can be found in many introductory textbooks (see e.g. [CK04; CCG+07; CCG+08; Bre10]).

Definition 5.1.1. Let M^n, $n \geq 2$ be a manifold. A curve of metrics $g(t)$ is called Ricci flow if it is a solution of the initial value problem

$$\frac{d}{dt}g(t) = -2\mathrm{Ric}_{g(t)}, \qquad g(0) = g_0. \tag{5.1}$$

It is well known that for a given metric g_0, there exists a short time interval $[0, \epsilon)$ and a Ricci flow $[0, \epsilon) \ni t \to g(t)$ starting at $g(0) = g_0$. Observe that the Ricci flow starting at an Einstein metric g_0 with constant μ is given by $(1-2t\mu)g_0$. So in the positive case, the manifold shrinks till it collapses at time $t = \frac{1}{2\mu}$. In the negative case, it expands for all time. Ricci-flat metrics remain unchanged under the flow. The Ricci flow is not a gradient flow in the strict sense but it can be interpreted as a gradient flow of the functional

$$\lambda(g) = \inf_{\substack{f \in C^\infty(M) \\ \int_M e^{-f}\, dV_g = 1}} \int_M (\mathrm{scal}_g + |\nabla f|_g^2) e^{-f}\, dV_g \tag{5.2}$$

on the space of metrics modulo diffeomorphisms. In fact, the first variation of λ is given by

$$\lambda(g)'(h) = -\int_M \langle h, \mathrm{Ric}_g + \nabla^2 f_g \rangle_g e^{-f_g}\, dV_g, \tag{5.3}$$

where f_g is the minimizer realizing $\lambda(g)$. Since λ is diffeomorphism invariant, its first variation vanishes on the space of Lie derivatives. In particular,

$$\lambda(g)'(\nabla^2 f_g) = \frac{1}{2}\lambda(g)'(\mathcal{L}_{\mathrm{grad} f_g} g) = 0.$$

Therefore,

$$\lambda(g)'(-2\mathrm{Ric}_g) = 2\int_M |\mathrm{Ric}_g + \nabla^2 f_g|_g^2 e^{-f_g}\, dV_g \geq 0,$$

which shows that λ is nondecreasing along the Ricci flow.

Remark 5.1.2. By substituting $\omega^2 = e^{-f}$, (5.2) becomes

$$\lambda(g) = \inf_{\substack{\omega \in C^\infty(M) \\ \int_M \omega^2\, dV_g = 1}} \int_M (\mathrm{scal}_g \omega^2 + 4|\nabla\omega|_g^2)\, dV_g. \tag{5.4}$$

This shows that $\lambda(g)$ is nothing but the smallest eigenvalue of the elliptic operator $H_g = 4\Delta_g + \mathrm{scal}_g$. In particular, if the scalar curvature of g is constant, we have $\lambda(g) = \mathrm{scal}_g$ and the corresponding minimizer satisfies $\omega \equiv \mathrm{vol}(M,g)^{-1/2}$.

As remarked above, the stationary points of the Ricci flow are precisely the Ricci-flat metrics and Einstein metrics remain stationary up to rescaling. It is therefore natural to ask how the Ricci flow behaves close to Einstein metrics. This question was discussed in the Ricci-flat case in [GIK02; Ses06; Has12; HM13] whereas the general Einstein case was discussed in [Ye93].

By work of Sesum and Haslhofer, the following was shown:

Theorem 5.1.3 ([Ses06],[Has12]). *Let (M, g_{RF}) be a compact Ricci-flat metric and suppose, all infinitesimal Einstein deformations are integrable. Then the follwing are equivalent:*

(i) For every neighbourhood $\mathcal{V}$ of g_{RF} in the space of metrics there exists a smaller neighbourhood $\mathcal{U} \subset \mathcal{V}$ such that the Ricci flow starting in $\mathcal{U}$ stays in $\mathcal{V}$ for all $t \geq 0$ and converges to a Ricci-flat metric for $t \to \infty$.

(ii) The Einstein operator is nonnegative on TT-tensors.

We call property (i) dynamical stability and (ii) linear stability. A central tool in Haslhofer's proof of this theorem is the λ-functional and its behavior near Ricci-flat metrics. In [Has12], an instability assertion is included: If neither the above cases of above do occur, then there exists an ancient (i.e. it exists since $t = -\infty$) Ricci flow $g(t)$, $t \in (-\infty, 0]$ such that $g(t) \to g_{RF}$ as $t \to -\infty$.

Recently, Haslhofer and Müller were able to get rid of the integrability assumption.

Theorem 5.1.4 ([HM13]). *Let (M, g_{RF}) be a compact Ricci-flat manifold. If g_{RF} is a local maximizer of λ, then for every $C^{k,\alpha}$-neighbourhood $\mathcal{U}$ of g_{RF} there exists a $C^{k,\alpha}$-neighbourhood $\mathcal{V}$ such that the Ricci flow starting at any metric in $\mathcal{V}$ exists for all time and converges (modulo diffeomorphism) to a Ricci-flat metric in $\mathcal{U}$.*

As above, [HM13] also includes an instability assertion: If the above assumption does not hold, there exists an ancient Ricci flow $g(t)$, $t \in (-\infty, 0]$ which converges modulo diffeomorphism to g_{RF} as $t \to -\infty$.

Remark 5.1.5. The maximality of λ can be characterized by the assertion that there exist no metrics of positive scalar curvature close to g_{RF} (see [CHI04, p. 4]). It is also equivalent to say that g_{RF} is a local maximum of the Yamabe functional. This is for example the case on any compact flat manifold and on the $K3$-surface.

Here, we give two classes of examples where this is the case.

Example 5.1.6 (Ricci-flat Kähler manifolds)**.** By [LeB99][Theorem 3.6], any compact four-dimensional Ricci-flat Kähler manifold (M, J, g) is supreme, i.e. it realizes the Yamabe invariant of M.

Example 5.1.7 (Manifolds with parallel spinors)**.** By [DWW05], any compact simply-connected manifold admitting a parallel spinor is a local maximum of the Yamabe functional.

Our aim now is to characterize dynamical stability for general Einstein metrics in the spirit of [Has12; HM13]. In this context, we consider dynamical stability with respect to certain normalized variants of the Ricci flow which leave Einstein metrics unchanged. From the results in [Ye93], dynamical stability follows if the Einstein operator Δ_E is positive on traceless tensors. We will prove dynamical stability under weaker assumptions and we will also prove instability theorems.

We will deal with functionals which are nondecreasing under these normalized Ricci flows. They are called Ricci entropies. It turns out that we have to use different functionals for positive and negative Einstein manifolds. Therefore we will treat both cases separately, although the strategy is basically the same.

In this chapter, we consider the Ricci flow close to negative Einstein manifolds. Without loss of generality, we may restrict to the case where the Einstein constant is equal to -1. Such metrics are stationary points of the flow

$$\dot{g}(t) = -2(\mathrm{Ric}_{g(t)} + g(t)). \tag{5.5}$$

This flow is homothetically equivalent to the standard Ricci flow. In fact,

$$\tilde{g}(t) = e^{-2t} g\left(\frac{1}{2}(e^{2t} - 1)\right)$$

is a solution of (5.5) starting at g_0 if and only if $g(t)$ is a solution of (5.1) starting at g_0.

Definition 5.1.8 (Dynamical stability and instability)**.** Let (M, g_E) be an Einstein manifold with constant -1. We call (M, g_E) dynamically stable if for every neighbourhood $\mathcal{U}$ of g_E in the space of metrics there exists a smaller neighbourhood $\mathcal{V} \subset \mathcal{U}$ such that the Ricci flow (5.5) starting in $\mathcal{V}$ stays in $\mathcal{U}$ for all $t \geq 0$ and converges to an Einstein metric with constant -1 for $t \to \infty$.

We call (M, g_E) dynamically stable modulo diffeomorphism if for each solution of (5.5) starting in $\mathcal{V}$, there exists a family of diffeomorphisms φ_t, $t \geq 0$ such that the modified flow $\varphi_t^* g(t)$ stays in $\mathcal{U}$ for all $t \geq 0$ and converges to an Einstein metric with constant -1 for $t \to \infty$.

We call (M, g_E) dynamically unstable (modulo diffeomorphism) if there exists an ancient Ricci flow $g(t)$, $t \in (-\infty, T]$ such that $g(t) \to g_E$ as $t \to -\infty$ (there exists a family of diffeomorphisms φ_t, $t \in (-\infty, T]$ such that $\varphi_t^* g(t) \to g_E$ as $t \to -\infty$).

Furthermore, from now on we call an Einstein manifold (M, g_E) Einstein-Hilbert stable if it is stable in the sense of Definition 2.5.1, i.e. the Einstein operator is nonnegative on TT-tensors. If this is not the case, (M, g_E) is called Einstein-Hilbert unstable.

5.2 The Expander Entropy

Let (M, g) be a Riemannian manifold and $f \in C^\infty(M)$. Define

$$\mathcal{W}_+(g,f) = \int_M \left[\frac{1}{2}(|\nabla f|^2 + \mathrm{scal}) - f\right] e^{-f}\, dV.$$

This is a simpler variant of the expander entropy $\mathcal{W}_+(g,f,\sigma)$ introduced in [FIN05].

Lemma 5.2.1. *The first variation of* $\mathcal{W}_+$ *at a tuple* (g,f) *equals*

$$\begin{aligned}\mathcal{W}_+'(h,v) = \int_M [-\frac{1}{2}\langle \mathrm{Ric} + \nabla^2 f - (-\Delta f - \frac{1}{2}|\nabla f|^2 + \frac{1}{2}\mathrm{scal} - f)g, h\rangle \\ - (-\Delta f - \frac{1}{2}|\nabla f|^2 + \frac{1}{2}\mathrm{scal} - f + 1)v]e^{-f}\, dV.\end{aligned}$$

Proof. Let $g_t = g + th$ and $f_t = f + tv$. We have

$$\begin{aligned}\frac{d}{dt}|_{t=0}\mathcal{W}_+(g_t,f_t) &= \int_M [\frac{1}{2}(|\nabla f_t|^2_{g_t} + \mathrm{scal}_{g_t}) - f_t]' e^{-f}\, dV \\ &+ \int_M [\frac{1}{2}(|\nabla f|^2 + \mathrm{scal}) - f](-v + \frac{1}{2}\mathrm{tr}h)e^{-f}\, dV.\end{aligned}$$

By the variational formula of the scalar curvature,

$$\begin{aligned}\int_M [\frac{1}{2}(|\nabla f_t|^2_{g_t} + \mathrm{scal}_{g_t}) - f_t]' e^{-f}\, dV \\ = \int_M (-\frac{1}{2}\langle h, \nabla f \otimes \nabla f\rangle + \langle \nabla f, \nabla v\rangle)e^{-f}\, dV \\ + \int_M [\frac{1}{2}(\Delta \mathrm{tr}h + \delta(\delta h) - \langle \mathrm{Ric}, h\rangle) - v]e^{-f}\, dV.\end{aligned}$$

By integration by parts,

$$\int_M \langle \nabla f, \nabla v\rangle e^{-f}\, dV = \int_M (\Delta f + |\nabla f|^2)ve^{-f}\, dV$$

and

$$\begin{aligned}\int_M \frac{1}{2}(\Delta \mathrm{tr}h + \delta(\delta h))e^{-f}\, dV &= \int_M \frac{1}{2}[\mathrm{tr}h\Delta(e^{-f}) + \langle h, \nabla^2(e^{-f})\rangle]\, dV \\ &= \int_M \frac{1}{2}[\mathrm{tr}h(-\Delta f - |\nabla f|^2) + \langle h, -\nabla^2 f + \nabla f \otimes \nabla f\rangle]e^{-f}\, dV.\end{aligned}$$

Thus,

$$\begin{aligned}\int_M [\frac{1}{2}(|\nabla f_t|^2_{g_t} + \mathrm{scal}_{g_t}) - f_t]' e^{-f}\, dV = \int_M [-\frac{1}{2}\langle h, \nabla^2 f + \mathrm{Ric} + (\Delta f + |\nabla f|^2)g\rangle \\ + (\Delta f + |\nabla f|^2 - 1)v]e^{-f}\, dV.\end{aligned}$$

The second term of above can be written as

$$\begin{aligned}\int_M [\frac{1}{2}(|\nabla f|^2 + \mathrm{scal}) - f](-v + \frac{1}{2}\mathrm{tr}h)e^{-f}\, dV \\ = \int_M [\frac{1}{2}\langle [\frac{1}{2}(|\nabla f|^2 + \mathrm{scal}) - f]g, h\rangle - [\frac{1}{2}(|\nabla f|^2 + \mathrm{scal}) - f]v]e^{-f}\, dV.\end{aligned}$$

By adding up these two terms, we obtain the desired formula. □

Now we introduce the functional

$$\mu_+(g) = \inf\left\{\mathcal{W}_+(g,f) \,\middle|\, f \in C^\infty(M),\ \int_M e^{-f}\, dV = 1\right\}. \tag{5.6}$$

It was shown in [FIN05, Thm 1.7] that given any smooth metric, the infimum is always uniquely realized by a smooth function. We call the minimizer f_g. The minimizer depends smoothly on the metric. It satisfies the Euler-Lagrange equation

$$-\Delta f_g - \frac{1}{2}|\nabla f_g|^2 + \frac{1}{2}\mathrm{scal}_g - f_g = \mu_+(g). \tag{5.7}$$

This can be seen as follows: If f_g realizes the infimum, then $\mathcal{W}_+'(0,v) = 0$ for all $v \in C^\infty(M)$ with $(v, e^{-f_g})_{L^2} = 0$ because of the constraint $\int_M e^{-f_g}\, dV = 1$. This is exactly the case when $-\Delta f_g - \frac{1}{2}|\nabla f_g|^2 + \frac{1}{2}\mathrm{scal}_g - f_g = c$ for some $c \in \mathbb{R}$. By integration by parts, one shows that

$$c = \int_M \left(-\Delta f_g - \frac{1}{2}|\nabla f_g|^2 + \frac{1}{2}\mathrm{scal} - f_g\right) e^{-f_g}\, dV = \mathcal{W}_+(g, f_g) = \mu_+(g). \tag{5.8}$$

Remark 5.2.2. Since $\mathcal{W}_+(\varphi^* g, \varphi^* f) = \mathcal{W}_+(g,f)$, the functional $\mu_+(g)$ is invariant under diffeomorphisms, so $\mu_+(\varphi^* g) = \mu_+(g)$ for any $\varphi \in \mathrm{Diff}(M)$.

Lemma 5.2.3 (First variation of μ_+). *The first variation of $\mu_+(g)$ is given by*

$$\mu_+(g)'(h) = -\frac{1}{2}\int_M \langle \mathrm{Ric} + g + \nabla^2 f_g, h\rangle e^{-f_g}\, dV, \tag{5.9}$$

where f_g realizes $\mu_+(g)$. As a consequence, μ_+ is nondecreasing under the Ricci flow (5.5).

Proof. By Lemma 5.2.1 and (5.7),

$$\mu_+(g)'(h) = \int_M \left[-\frac{1}{2}\langle \mathrm{Ric} + \nabla^2 f - \mu_+(g) g, h\rangle - (\mu_+(g)+1)v\right] e^{-f}\, dV \tag{5.10}$$

where $v = \frac{d}{dt}|_{t=0} f_{g+th}$. Due to the constraint $\int_M e^{-f_g}\, dV_g = 1$, we have $(v, e^{-f_g})_{L^2} = \frac{1}{2}\int_M \mathrm{tr} h\, dV$. Inserting this in (5.10) yields the first variational formula. By diffeomorphism invariance,

$$\mu_+(g)'(\nabla^2 f_g) = \frac{1}{2}\mu_+'(g)(\mathcal{L}_{\mathrm{grad} f_g} g) = 0.$$

Thus, if $g(t)$ is a solution of (5.5),

$$\frac{d}{dt}\mu_+(g(t)) = \int_M |\mathrm{Ric}_{g(t)} + g(t) + \nabla^2 f_{g(t)}|^2 e^{-f_{g(t)}}\, dV_{g(t)} \geq 0,$$

which proves the lemma. □

Remark 5.2.4. We call metrics gradient Ricci solitons if $\mathrm{Ric}_g + \nabla^2 f = cg$ for some $f \in C^\infty(M)$ and $c \in \mathbb{R}$. In the compact case, any such metric is already Einstein if $c \leq 0$ (see [Cao10, Proposition 1.1]). By the first variational formula of μ_+, we conclude that Einstein metrics with constant -1 are precisely the critical points of μ_+.

Lemma 5.2.5. *Let (M, g_E) be an Einstein manifold with constant -1. Furthermore, let $h \in \delta_{g_E}^{-1}(0)$. Then*

(i) $f_{g_E} \equiv \log \mathrm{vol}(M, g_E)$,

(ii) $\frac{d}{dt}|_{t=0} f_{g_E+th} = \frac{1}{2}\mathrm{tr}_{g_E} h$,

(iii) $\frac{d}{dt}|_{t=0}(\mathrm{Ric}_{g_E+th} + g_E + th + \nabla^2_{g_E+th} f_{g_E+th}) = \frac{1}{2}\Delta_E h$,

where Δ_E is the Einstein operator.

Proof. By substituting $w = e^{-f/2}$, we see that $w_{g_E} = e^{-f_{g_E}/2}$ is the minimizer of the functional

$$\widetilde{\mathcal{W}}(w) = \int_M 2|\nabla w|^2 + \frac{1}{2}\mathrm{scal} w^2 + w^2 \log w^2 \, dV$$

under the constraint

$$\|w\|_{L^2} = 1.$$

By Jensen's inequality, we have a lower bound

$$\widetilde{\mathcal{W}}(w) \geq \frac{1}{2} \inf_{p \in M} \mathrm{scal}(p) - \log(\mathrm{vol}(M, g_E)), \tag{5.11}$$

which is realized by the constant function $w_{g_E} \equiv \mathrm{vol}(M, g_E)^{-1/2}$ since the scalar curvature is constant on M. This proves (i).

To prove (ii), we differentiate the Euler-Lagrange equation (5.7) in the direction of h. We obtain

$$\begin{aligned} 0 = (-\Delta f)' - \frac{1}{2}(|\nabla f|^2)' + \frac{1}{2}\mathrm{scal}' - f' &= -\Delta f' + \frac{1}{2}(\Delta(\mathrm{tr} h) + \delta(\delta h) + \mathrm{tr} h) - f' \\ &= -(\Delta + 1) f' + \frac{1}{2}(\Delta + 1)\mathrm{tr} h. \end{aligned}$$

Here we used that f_{g_E} is constant and $\delta h = 0$. Since $\Delta + 1$ is invertible on the space of smooth functions, the second assertion follows. It remains to show (iii). By straightforward differentiation,

$$\begin{aligned} (\mathrm{Ric} + g + \nabla^2 f)' &= \frac{1}{2}\Delta_L h - \delta^*(\delta h) - \frac{1}{2}\nabla^2 \mathrm{tr} h + h + (\nabla^2)' f_{g_E} + \nabla^2(f') \\ &= \frac{1}{2}\Delta_L h - \frac{1}{2}\nabla^2 \mathrm{tr} h + h + \frac{1}{2}\nabla^2 \mathrm{tr} h = \frac{1}{2}\Delta_E h. \end{aligned}$$

Here we used (i) and (ii). □

Proposition 5.2.6 (Second variation of μ_+). *The second variation of μ_+ at an Einstein metric with $\mathrm{Ric}_{g_E} = -g_E$ is given by*

$$\mu_+(g_E)''(h) = \begin{cases} -\frac{1}{4} \fint_M \langle \Delta_E h, h \rangle \, dV, & \text{if } h \in \delta^{-1}(0), \\ 0, & \text{if } h \in \delta^*(\Omega^1(M)), \end{cases}$$

where $\fint$ denotes the averaging integral, i.e. the integral divided by the volume.

Proof. Since μ_+ is a Riemannian functional, the Hessian restricted to $\delta^*(\Omega^1(M))$ vanishes. Now let $h \in \delta^{-1}(0)$. By the first variational formula and Lemma 5.2.5,

$$\begin{aligned}\mu_+(g_E)''(h) &= \left.\frac{d^2}{dt^2}\right|_{t=0} \mu_+(g_E + th) \\ &= -\frac{1}{2}\int_M \langle (\mathrm{Ric} + g + \nabla^2 f)', h\rangle e^{-f}\, dV \\ &= -\frac{1}{4}\fint_M \langle \Delta_E h, h\rangle\, dV.\end{aligned}$$

Recall that Δ_E preserves $\delta^{-1}(0)$. Thus, the splitting $\delta^*(\Omega^1(M)) \oplus \delta^{-1}(0)$ is orthogonal with respect to μ_+''. □

With this formula, we easily prove

Corollary 5.2.7. *Let (M, g_E) be an Einstein manifold with constant -1. Then dynamical stability (modulo diffeomorphism) implies Einstein-Hilbert stability.*

Proof. We have seen that μ_+ is nondecreasing under (5.5) and that μ_+ is invariant under diffeomorphisms. Thus, (M, g_E) is nessecarily a local maximum of μ_+, if it is dynamically stable (modulo diffeomorphism). Consequently, the second variation of μ_+ is negative semidefinite. The assertion now follows from Proposition 5.2.6. □

5.3 Some technical Estimates

In this section, we will establish bounds on μ_+, f_g and their variations in terms of certain norms of the variations. These estimates are needed in proving the main theorems of the next two sections.

Lemma 5.3.1. *Let (M, g_E) be an Einstein manifold such that $\mathrm{Ric} = -g_E$. Then there exists a $C^{2,\alpha}$-neighbourhood $\mathcal{U}$ in the space of metrics such that the minimizers f_g are uniformly bounded in $C^{2,\alpha}$, i.e. there exists a constant $C > 0$ such that $\|f_g\|_{C^{2,\alpha}} \leq C$ for all $g \in \mathcal{U}$. Moreover, for each $\epsilon > 0$, we can choose $\mathcal{U}$ so small that $\|\nabla f_g\|_{C^0} \leq \epsilon$ for all $g \in \mathcal{U}$.*

Proof. As in the proof of Lemma 5.2.5 (i), we use the fact that

$$\mu_+(g) = \inf_{w \in C^\infty(M)} \widetilde{\mathcal{W}}(g, w) = \inf \int_M 2|\nabla w|^2 + \frac{1}{2}\mathrm{scal} w^2 + w^2 \log w^2\, dV \quad (5.12)$$

under the constraint

$$\|w\|_{L^2} = 1.$$

There exists a unique minimizer of this functional which we denote by w_g. We have $w_g = e^{-f_g/2}$ and w_g satisfies the Euler-Lagrange equation

$$2\Delta w_g + \frac{1}{2}\mathrm{scal}_g w_g - 2 w_g \log w_g = \mu_+(g) w_g. \quad (5.13)$$

We will now show that there exists a uniform bound $\|w_g\|_{C^{2,\alpha}} \leq C$ for all metrics g in a $C^{2,\alpha}$-neighbourhood $\mathcal{U}$ of g_E. First observe that by (5.12),

$$2\|\nabla w_g\|_{L^2} \leq \mu_+(g) - C_1 \mathrm{vol}(M, g) - \frac{1}{2}\inf_{p \in M} \mathrm{scal}_g(p),$$

since the function $x \mapsto x \log x$ has a lower bound. Testing $\widetilde{\mathcal{W}}$ with the constant function $w \equiv \mathrm{vol}(M, gs)^{-1/2}$ yields

$$\mu_+(g) \leq \frac{1}{2} \sup_{p \in M} \mathrm{scal}_g(p) - \log(\mathrm{vol}(M, g)). \tag{5.14}$$

Using these estimates, we see that the H^1-norm of ω_g can be estimated by quantities, which are uniformly bounded on $\mathcal{U}$. By Sobolev embedding, we obtain a uniform bound on $\|w_g\|_{L^{2n/(n-2)}}$. Let $p = 2n/(n-2)$ and choose some q slightly smaller than p. By (5.13) and elliptic regularity,

$$\|w_g\|_{W^{2,q}} \leq C_2(\|w_g \log w_g\|_{L^q} + \|w_g\|_{L^q}).$$

Since $x \mapsto x \log x$ grows slower than $x \mapsto x^\beta$ for any $\beta > 1$ as $x \to \infty$, we have the estimate

$$\|w_g \log w_g\|_{L^q} \leq C_3(\mathrm{vol}(M, g)) + \|w_g\|_{L^p}.$$

This yields an uniform bound $\|w_g\|_{W^{2,q}} \leq C(q)$. By Sobolev embedding, we have uniform bounds on $\|w_g\|_{L^{p'}}$ for some $p' > p$ and by applying elliptic regularity on (5.13), we have bounds on $\|w_g\|_{W^{2,q'}}$ for every $q' < p'$.

Iterating this procedure, we obtain uniform bounds $\|w_g\|_{W^{2,p}} \leq C(p)$ for each $p \in (1, \infty)$. Again by elliptic regularity ,

$$\begin{aligned} \|w_g\|_{C^{2,\alpha}} &\leq C_4(\|w_g \log w_g\|_{C^{0,\alpha}} + \|w_g\|_{C^{0,\alpha}}) \\ &\leq C_5[(\|w_g\|_{C^{0,\alpha}})^\gamma + \|w_g\|_{C^{0,\alpha}}) \end{aligned}$$

for some $\gamma > 1$ and for sufficiently large p, Sobolev embedding yields

$$\|w_g\|_{C^{0,\alpha}} \leq C_6 \|w_g\|_{W^{1,p}} \leq C_6 \cdot C(p).$$

Therefore, we have a uniform bound on $\|w_g\|_{C^{2,\alpha}}$.

Next, we show that the $C^{2,\alpha}$-norms of f_g are uniformly bounded. First, we claim that we may choose a smaller neighbourhood $\mathcal{V} \subset \mathcal{U}$ such that for $g \in \mathcal{V}$, the functions w_g are bounded away from zero (recall that any $w_g = e^{-f_g/2}$ is positive). Suppose this is not the case. Then there exists a sequence $g_i \to g_E$ in $C^{2,\alpha}$ such that $\min_p w_{g_i}(p) \to 0$ for $i \to \infty$. Since $\|w_{g_i}\|_{C^{2,\alpha}} \leq C$ for all i, there exists a subsequence, again denoted by w_{g_i} such that $w_{g_i} \to w_\infty$ in $C^{2,\alpha'}$ for some $\alpha' < \alpha$. Obviously, the right hand side of (5.12) converges. By the estimates (5.11) and (5.14), we see that also the left hand side of (5.12) converges. Therefore, w_∞ equals the minimizer of $\widetilde{\mathcal{W}}(g_E, w)$, so $w_\infty = w_{g_E} = \mathrm{vol}(M, g_E)^{-1/2}$. In particular, $\min_p w_{g_i}(p) \to \mathrm{vol}(M, g_E)^{-1/2} \neq 0$ which contradicts the assumption. Now we have

$$\begin{aligned} \|f_g\|_{C^{2,\alpha}} = \|-2\log(w_g)\|_{C^{2,\alpha}} &\leq C(\log \inf w_g, 1/(\inf w_g)) \|w_g\|_{C^{2,\alpha}} \\ &\leq C(\log \inf w_g, 1/(\inf w_g)) \cdot C_6 \cdot C(p) \end{aligned}$$

and the claim shows that the right hand side is bounded.

It remains to prove that for each $\epsilon > 0$, we may choose $\mathcal{U}$ so small that $\|\nabla f_g\|_{C^0} < \epsilon$. We again use a subsequence argument. Suppose this is not possible. Then there exists a sequence of metrics $g_i \to g$ in $C^{2,\alpha}$ and some $\epsilon_0 > 0$ such that for the corresponding f_{g_i}, the estimate $\|\nabla f_{g_i}\|_{C^0} \geq \epsilon_0$ holds for all i. Because of the bound $\|f_g\|_{C^{2,\alpha}} \leq C$, we may choose a subsequence, again denoted by f_i converging to some f_∞ in $C^{2,\alpha'}$ for $\alpha' < \alpha$. By the same arguments as above, $f_\infty = f_{g_E} \equiv -\log(\mathrm{vol}(M))$. In particular, $\|\nabla f_{g_i}\|_{C^0} \to 0$, a contradiction. $\square$

Lemma 5.3.2. *Let (M, g_E) be an Einstein manifold such that $\mathrm{Ric}_{g_E} = -g_E$. Then there exists a $C^{2,\alpha}$-neighbourhood $\mathcal{U}$ of g_E in the space of metrics and a constant $C > 0$ such that for all $g \in \mathcal{U}$, we have*

$$\left\| \frac{d}{dt}\Big|_{t=0} f_{g+th} \right\|_{C^{2,\alpha}} \leq C \|h\|_{C^{2,\alpha}}, \qquad \left\| \frac{d}{dt}\Big|_{t=0} f_{g+th} \right\|_{H^i} \leq C \|h\|_{H^i}, \quad i = 1, 2.$$

Proof. Recall that f_g satisfies the Euler-Lagrange equation

$$-\Delta f - \frac{1}{2}|\nabla f|^2 + \frac{1}{2}\mathrm{scal} - f = \mu_+(g).$$

Differentiating this equation in the direction of h yields

$$-\dot{\Delta} f - \Delta \dot{f} + \frac{1}{2}h(\mathrm{grad}f, \mathrm{grad}f) - \langle \nabla f, \nabla \dot{f} \rangle + \frac{1}{2}\dot{\mathrm{scal}} - \dot{f} = \dot{\mu}_+(g).$$

By Lemma A.3 and Lemma A.2 the variational formulas for the Laplacian and the scalar curvature are

$$\dot{\Delta} f = \langle h, \nabla^2 f \rangle - \langle \delta h + \frac{1}{2}\nabla \mathrm{tr} h, \nabla f \rangle,$$
$$\dot{\mathrm{scal}} = \Delta(\mathrm{tr} h) + \delta(\delta h) - \langle \mathrm{Ric}, h \rangle.$$

Because $\Delta + 1$ is invertible, we can apply elliptic regularity and we obtain

$$\begin{aligned}
&\left\|\dot{f}\right\|_{C^{2,\alpha}} \leq C_1 \left\|(\Delta + 1)\dot{f}\right\|_{C^{0,\alpha}} \\
&\leq C_1 \left\|\langle \nabla f, \nabla \dot{f} \rangle\right\|_{C^{0,\alpha}} + C_1 \left\| -\dot{\Delta} f + \frac{1}{2}h(\nabla f, \nabla f) + \frac{1}{2}\dot{\mathrm{scal}} - \dot{\mu}_+(g) \right\|_{C^{0,\alpha}} \\
&\leq C_1 \|\nabla f\|_{C^0} \left\|\nabla \dot{f}\right\|_{C^{0,\alpha}} + C_1 \left\| -\dot{\Delta} f + \frac{1}{2}h(\nabla f, \nabla f) + \frac{1}{2}\dot{\mathrm{scal}} - \dot{\mu}_+(g) \right\|_{C^{0,\alpha}}.
\end{aligned}$$

By Lemma 5.3.1, we may choose $\mathcal{U}$ so small that $\|\nabla f\|_{C^0} < \epsilon$ for some small $\epsilon < \min\{C_1^{-1}, 1\}$. Then we have

$$\begin{aligned}
(1 - \epsilon) \left\|\dot{f}\right\|_{C^{2,\alpha}} &\leq \left\|\dot{f}\right\|_{C^{2,\alpha}} - C_1 \|\nabla f\|_{C^0} \left\|\nabla \dot{f}\right\|_{C^{0,\alpha}} \\
&\leq C_1 \left\| -\dot{\Delta} f + \frac{1}{2}h(\nabla f, \nabla f) + \frac{1}{2}\dot{\mathrm{scal}} - \dot{\mu}_+(g) \right\|_{C^{0,\alpha}} \\
&\leq (C_2 \|f_g\|_{C^{2,\alpha}} + C_3) \|h\|_{C^{2,\alpha}}.
\end{aligned}$$

The last inequality follows from the variational formulas of the Laplacian, the scalar curvature and μ_+. By the uniform bound on $\|f_g\|_{C^{2,\alpha}}$, the first estimate of the lemma follows.
Similarly, we can estimate the H^i-norm of $\dot{f}$. Again by elliptic regularity,

$$\begin{aligned}
&\left\|\dot{f}\right\|_{H^i} \leq C_4 \left\|(\Delta + 1)\dot{f}\right\|_{H^{i-2}} \\
&\leq C_4 \left\|\langle \nabla f, \nabla \dot{f} \rangle\right\|_{L^2} + C_4 \left\| -\dot{\Delta} f + \frac{1}{2}h(\nabla f, \nabla f) + \frac{1}{2}\dot{\mathrm{scal}} - \dot{\mu}_+(g) \right\|_{H^{i-2}} \\
&\leq C_4 \|\nabla f\|_{C^0} \left\|\nabla \dot{f}\right\|_{L^2} + C_4 \left\| -\dot{\Delta} f + \frac{1}{2}h(\nabla f, \nabla f) + \frac{1}{2}\dot{\mathrm{scal}} - \dot{\mu}_+(g) \right\|_{H^{i-2}}.
\end{aligned}$$

Choosing $\mathcal{U}$ such that $\|\nabla f\|_{C^0} < \epsilon$ for some $\epsilon < \min\{C_4^{-1}, 1\}$, we obtain

$$\begin{aligned}(1-\epsilon)\left\|\dot{f}\right\|_{H^i} &\leq \left\|\dot{f}\right\|_{H^i} - C_4\|\nabla f\|_{C^0}\left\|\nabla \dot{f}\right\|_{L^2}\\ &\leq C_4\left\|-\dot{\Delta} f + \frac{1}{2}h(\nabla f,\nabla f) + \frac{1}{2}\dot{\text{scal}} - \dot{\mu}_+(g)\right\|_{H^{i-2}}\\ &\leq (C_5\|f\|_{C^{2,\alpha}} + C_6)\|h\|_{H^i}.\end{aligned}$$

The last estimate is clear from the variational formulas if $H^{i-2} = L^2$. In the case $H^{i-2} = H^{-1}$, we test the integrand with an H^1-function φ and find that

$$(-\dot{\Delta} f + \frac{1}{2}h(\nabla f,\nabla f) + \frac{1}{2}\dot{\text{scal}} - \dot{\mu}_+(g), \varphi)_{L^2} \leq (C_5\|f\|_{C^{2,\alpha}} + C_6)\|h\|_{H^1}\|\varphi\|_{H^1}.$$

Therefore, the H^{-1}-norm can be estimated as above. □

Proposition 5.3.3 (Estimate of the second variation of μ_+). *Let (M, g_E) be an Einstein manifold with constant -1. Then there exists a $C^{2,\alpha}$-neighbourhood $\mathcal{U}$ of g_E and a constant $C > 0$ such that*

$$\left|\left.\frac{d^2}{dsdt}\right|_{s,t=0}\mu_+(g+th+sk)\right| \leq C\|h\|_{H^1}\|k\|_{H^1}$$

for all $g \in \mathcal{U}$.

Proof. By the formula of the first variation,

$$\begin{aligned}\left.\frac{d^2}{dsdt}\right|_{s,t=0}\mu_+(g+th+sk) &= -\left.\frac{d}{ds}\right|_{s=0}\frac{1}{2}\int_M\langle \text{Ric}_{g_s} + g_s - \nabla^2 f_{g_s}, h\rangle_{g_s} e^{-f_{g_s}}\, dV_{g_s}\\ &= (1) + (2) + (3),\end{aligned}$$

and we estimate these three terms separately. The first term comes from differentiating the scalar product:

$$\begin{aligned}|(1)| &= \left|\int_M\langle \text{Ric}_g + g - \nabla^2 f_g, k\circ h\rangle_g e^{-f_g}\, dV_g\right|\\ &\leq C_1\left\|\text{Ric}_g + g - \nabla^2 f_g\right\|_{C^0}\left\|e^{-f_g}\right\|_{C^0}\|h\|_{L^2}\|k\|_{L^2}\\ &\leq (C_2 + C_3\|f\|_{C^{2,\alpha}})\exp(-\min f_g)\|h\|_{L^2}\|k\|_{L^2}\\ &\leq C_4\|h\|_{H^1}\|k\|_{H^1}.\end{aligned}$$

These estimates hold since $|\text{Ric}_g + g|$ and $\exp(-\min f_g)$ are uniformly bounded in a small $C^{2,\alpha}$-neighbourhood of g_E. The second term comes from differentiating the gradient:

$$\begin{aligned}|(2)| &= \left|\frac{1}{2}\int_M\left\langle\left.\frac{d}{ds}\right|_{s=0}(\text{Ric}_{g_s} + g_s - \nabla^2 f_{g_s}), h\right\rangle_g e^{-f_g}\, dV_g\right|\\ &= \left|\frac{1}{2}\int_M\langle \text{Ric}' + k - (\nabla^2)' f_g - \nabla^2 f_g', h\rangle_g e^{-f_g}\, dV_g\right|\\ &= \left|\frac{1}{2}\int_M\left\langle\frac{1}{2}\Delta_L k - \delta^*(\delta k) - \frac{1}{2}\nabla^2\text{tr}k + k - (\nabla^2)' f_g - \nabla^2 f_g', e^{-f_g}h\right\rangle_g dV_g\right|\\ &\leq C_5\|k\|_{H^1}\|h\|_{H^1}\|f_g\|_{C^{2,\alpha}}\exp(-\min f_g)\\ &\leq C_6\|k\|_{H^1}\|h\|_{H^1}.\end{aligned}$$

The first inequality follows from integration by parts, Lemma A.3 and Lemma 5.3.2, the second from the uniform bound on the f_g. The third term appears when we differentiate the measure:

$$\begin{aligned}|(3)| &= \left|\frac{1}{2}\int_M \langle \mathrm{Ric}_g + g - \nabla^2 f_g, h\rangle_g \left(-f_g' + \frac{1}{2}\mathrm{tr}k\right) e^{-f_g}\, dV_g\right| \\ &\leq C_7 \left\|\mathrm{Ric}_g + g - \nabla^2 f_g\right\|_{C^0} \exp(-\min f_g) \left\|h\right\|_{L^2} \left\|-f_g' + \frac{1}{2}\mathrm{tr}k\right\|_{L^2} \\ &\leq C_8 \left\|h\right\|_{H^1} \left\|k\right\|_{H^1}.\end{aligned}$$

Here, we again used Lemma 5.3.2 in the last step. □

Lemma 5.3.4. *Let (M, g_E) be an Einstein manifold with constant -1. Then there exists a $C^{2,\alpha}$-neighbourhood $\mathcal{U}$ of g_E and a constant $C > 0$ such that*

$$\left\|\left.\frac{d^2}{dtds}\right|_{t,s=0} f_{g+sk+th}\right\|_{H^i} \leq C \left\|h\right\|_{C^{2,\alpha}} \left\|k\right\|_{H^i}, \qquad i = 1, 2.$$

Proof. In the proof, we denote t-derivatives by dot and s-derivatives by prime. Differentiating (5.7) twice yields

$$\begin{aligned}-\Delta \dot{f}' - \dot{\Delta} f' - \Delta' \dot{f} - \dot{\Delta}' f + h(\mathrm{grad} f, \mathrm{grad} f') + k(\mathrm{grad} f, \mathrm{grad} \dot{f}) \\ -\langle \nabla f, \nabla \dot{f}'\rangle - \langle \nabla \dot{f}, \nabla f'\rangle + \frac{1}{2}\dot{\mathrm{scal}}' - \dot{f}' = \dot{\mu}_+'.\end{aligned}$$

By elliptic regularity, we have

$$\left\|\dot{f}'\right\|_{H^i} \leq C_1 \left\|(\Delta + 1)\dot{f}'\right\|_{H^{i-2}} \leq C_1 \left\|\nabla f\right\|_{C^0} \left\|\nabla \dot{f}'\right\|_{L^2} + C_1 \left\|(A)\right\|_{H^{i-2}}, \quad (5.15)$$

where

$$\begin{aligned}(A) = &- \dot{\Delta} f' - \Delta' \dot{f} - \dot{\Delta}' f + h(\mathrm{grad} f, \mathrm{grad} f') + k(\mathrm{grad} f, \mathrm{grad} \dot{f}) \\ &- \langle \nabla \dot{f}, \nabla f'\rangle + \frac{1}{2}\dot{\mathrm{scal}}' - \dot{\mu}_+'.\end{aligned}$$

Now we consider the occurent second variational formulas of the Laplacian and the scalar curvature. By Lemma A.5, they can be schematically written as

$$\begin{aligned}\dot{\Delta}' f &= \nabla k * h * \nabla f + k * \nabla h * \nabla f, \\ \dot{\mathrm{scal}}' &= \nabla^2 k * h + k * \nabla^2 h + \nabla k * \nabla h + R * k * h.\end{aligned}$$

Here, $*$ is Hamilton's notation for a combination of tensor products with contractions. Now, Lemma 5.3.3, integration by parts and the Hölder inequality yield

$$\left\|-\dot{\Delta}' f + \frac{1}{2}\dot{\mathrm{scal}}' - \dot{\mu}_+'\right\|_{H^{i-2}} \leq C_2 \left\|h\right\|_{C^{2,\alpha}} \left\|k\right\|_{H^i}.$$

For $H^{i-2} = L^2$, this is clear, for $H^{i-2} = H^{-1}$ this follows again from testing with an H^1-function. For the remaining terms

$$(B) = -\dot{\Delta} f' - \Delta' \dot{f} + h(\mathrm{grad} f, \mathrm{grad} f') + k(\mathrm{grad} f, \mathrm{grad} \dot{f}) - \langle \nabla \dot{f}, \nabla f'\rangle$$

such an estimate follows from the first variational formula of the Laplacian and the estimates we already developed for $\dot{f}$ and f' in Lemma 5.3.2. In fact, we even have

$$\|(B)\|_{L^2} \leq C_3 \|h\|_{C^{2,\alpha}} \|k\|_{H^1},$$

so that the desired type of estimate holds in any case. We obtain

$$\|(A)\|_{H^{i-2}} \leq C_4 \|h\|_{C^{2,\alpha}} \|k\|_{H^i}.$$

Since $\|\nabla f\|_{C^0}$ can be assumed to be arbitrarily small, we bring this term to the left hand side of (5.15) and obtain the result. □

Proposition 5.3.5 (Estimates of the third variation of μ_+). *Let (M, g_E) be an Einstein manifold with constant -1. Then there exists a $C^{2,\alpha}$-neighbourhood $\mathcal{U}$ of g_E and a constant $C > 0$ such that*

$$\left|\frac{d^3}{dt^3}\right|_{t=0} \mu_+(g+th)\Big| \leq C \|h\|_{H^1}^2 \|h\|_{C^{2,\alpha}}$$

for all $g \in \mathcal{U}$.

Proof. We have, by the first variational formula,

$$\begin{aligned}
\frac{d^3}{dt^3}\Big|_{t=0} \mu_+(g+th) &= -\frac{1}{2}\frac{d^2}{dt^2}\Big|_{t=0} \int_M \langle \mathrm{Ric} + g + \nabla^2 f_g, h\rangle e^{-f}\, dV \\
&= -\frac{1}{2}\int_M \langle (\mathrm{Ric} + g + \nabla^2 f_g)'', h\rangle e^{-f}\, dV - 3\int_M \langle \mathrm{Ric} + g + \nabla^2 f_g, h^3\rangle e^{-f}\, dV \\
&- \frac{1}{2}\int_M \langle \mathrm{Ric} + g + \nabla^2 f_g, h\rangle (e^{-f}\, dV)'' + 2\int_M \langle (\mathrm{Ric} + g + \nabla^2 f_g)', h\circ h\rangle e^{-f}\, dV \\
&- \int_M \langle (\mathrm{Ric} + g + \nabla^2 f_g)', h\rangle (e^{-f}\, dV)' + 2\int_M \langle \mathrm{Ric} + g + \nabla^2 f_g, h\circ h\rangle (e^{-f}\, dV)',
\end{aligned}$$

where $h^3 = h\circ h\circ h$. Straightforward calculations show that

$$\begin{aligned}
(e^{-f}\, dV)' &= \left(-f' + \frac{1}{2}\mathrm{tr}h\right) e^{-f}\, dV, \\
(e^{-f}\, dV)'' &= \left[-f'' - \frac{1}{2}|h|^2 + \left(-f' + \frac{1}{2}\mathrm{tr}h\right)^2\right] e^{-f}\, dV, \\
(\mathrm{Ric} + g + \nabla^2 f_g)' &= \frac{1}{2}\Delta_L h - \delta^*(\delta h) - \frac{1}{2}\nabla^2 \mathrm{tr}h + h - (\nabla^2)' f_g - \nabla^2 f_g'.
\end{aligned}$$

By these calculations, it is quite obvious that we can estimate the last five terms of above by $C\|h\|_{H^1}^2 \|h\|_{C^{2,\alpha}}$ using the Hölder inequality and the estimates for f' and f'' we developed in Lemma 5.3.2 and Lemma 5.3.4. It remains to deal with the first term which contains the second derivative of the gradient of μ_+. We have the schematic expressios

$$\begin{aligned}
(\mathrm{Ric} + g)'' &= \nabla^2 h * h + \nabla h * \nabla h + R * h * h, \\
(\nabla^2 f_g)'' &= (\nabla^2)'' f_g + 2(\nabla^2)' f_g' + \nabla^2 f_g'' \\
&= \nabla f * \nabla h * h + \nabla f' * \nabla h + \nabla^2 f_g'',
\end{aligned}$$

see Lemma A.3 and Lemma A.5. From these expressions we obtain, by applying Lemma 5.3.2 and Lemma 5.3.4 again,

$$\left| \int_M \langle (\mathrm{Ric} + g + \nabla^2 f_g)'', h \rangle e^{-f} \, dV \right| \leq C \|h\|_{H^1}^2 \|h\|_{C^{2,\alpha}} .$$

Note that we have to use integration by parts for the $\nabla^2 f_g''$-term to obtain an upper bound containing an H^1-norm. □

5.4 The integrable Case

In this section, we prove dynamical stability and instability theorems for negative Einstein manifolds under the technical assumption that all infinitesimal Einstein deformations are integrable. For all results in this section, we suppose that this condition holds.

5.4.1 Local Maximum of the Expander Entropy

In this section we will prove a criterion which ensures that an Einstein manifold is a local maximum of the expander entropy. For the proof, we will use the slice theorem stated in Chapter 2 and we use an explicit construction of the slice. Moreover, we use Taylor expansion and with the estimates of the previous section, we are able to control the error terms.

Theorem 5.4.1 (Ebin-Palais ([Ebi70])). *Let M be a compact manifold and $\mathcal{M}$ the space of metrics on M. Then for each $g_0 \in \mathcal{M}$, there exists a $C^{2,\alpha}$-neighbourhood $\mathcal{U}$ such that each $g \in \mathcal{U}$ can be written as $g = \varphi^* \tilde{g}$ for some $\varphi \in \mathrm{Diff}(M)$ and $\tilde{g} \in \mathcal{S}_{g_0} = (g_0 + \delta_{g_0}^{-1}(0)) \cap \mathcal{U}$.*

We call S_{g_0} an affine slice. Now let g_E be an Einstein metric with constant -1 and let

$$\mathcal{E} = \{ g \in \mathcal{S}_{g_E} \mid \mathrm{Ric}_g = \alpha g \text{ for some } \alpha \in \mathbb{R} \}$$

be the set of Einstein metrics in the affine slice near g_E. Moreover, let

$$\mathcal{P} = \{ g \in \mathcal{S}_{g_E} \mid \mathrm{Ric}_g = -g \} .$$

If we assume that all infinitesimal Einstein deformations of g_E are integrable, $\mathcal{E}$ (and hence also $\mathcal{P}$) is a manifold near g_E and the tangent spaces at g_E are given by

$$T_{g_E}\mathcal{E} = \mathbb{R} \cdot g_E \oplus \ker(\Delta_E|_{TT}), \qquad T_{g_E}\mathcal{P} = \ker(\Delta_E|_{TT}),$$

see [Bes08, Proposition 12.49] for more details. Note also that

$$\mathcal{P} = \{ g \in \mathcal{E} \mid \mathrm{vol}(M, g) = \mathrm{vol}(M, g_E) \}$$

by the observations in Section 2.5. By Lemma 5.2.5 (i), f_g is constant for any $g \in \mathcal{P}$ and thus, $\mu_+(g) = \frac{\mathrm{scal}_g}{2} - \log(\mathrm{vol}(M, g))$ is constant on $\mathcal{P}$. For the proof of maximality, we further need the following decomposition of the space of divergence-free tensors:

Lemma 5.4.2. *Let (M, g_E) be an Einstein manifold with nonvanishing constant μ. Then we have the L^2-orthogonal decomposition*

$$\delta_{g_E}^{-1}(0) = C_{g_E}(C^\infty(M)) \oplus TT_{g_E},$$

where $C_{g_E} : C^\infty(M) \to \Gamma(S^2M)$ is defined as $C_{g_E}(f) = (\Delta f - \mu f)g_E + \nabla^2 f$.

Proof. We first check the L^2-orthgonality. Let $f \in C^\infty(M)$ and $h \in TT_{g_E}$. Then

$$(C_{g_E} f, h)_{L^2} = (\Delta f - \mu f, \mathrm{tr} h)_{L^2} + (\nabla^2 f, h)_{L^2} = 0 + (\nabla f, \delta h)_{L^2} = 0.$$

Now we show that $C_{g_E}(C^\infty(M)) \subset \delta_{g_E}^{-1}(0)$. Let $f \in C^\infty(M)$. We then have

$$\begin{aligned}\delta(C_{g_E} f) =&\delta((\Delta f - \mu f)g_E + \nabla^2 f)\\ =&-\nabla\Delta f + \mu\nabla f + \delta\nabla^2 f.\end{aligned}$$

Let $\{e_1, \ldots, e_n\}$ be a local orthonormal frame. By the Ricci identity,

$$\begin{aligned}\delta(C_{g_E} f)(e_i) =& -\nabla_{e_i}\Delta f + \mu\nabla_{e_i} f + \delta\nabla^2 f(e_i)\\ =&\sum_{j=1}^n (\nabla^3_{e_i,e_j,e_j} - \nabla^3_{e_j,e_i,e_j})f + \mu\nabla_{e_i} f\\ =&\sum_{j=1}^n \nabla_{R(e_j,e_i)e_j} f + \mu\nabla_{e_i} f\\ =&-\mu\nabla_{e_i} f + \mu\nabla_{e_i} f = 0.\end{aligned}$$

Since also $TT_{g_E} \subset \delta_{g_E}^{-1}(0)$ and the decomposition $\Gamma(S^2M) = \delta^*(\Omega^1(M)) \oplus \delta_{g_E}^{-1}(0)$ is orthogonal, it suffices to prove

$$\Gamma(S^2M) = \delta^*(\Omega^1(M)) \oplus C_{g_E}(C^\infty(M)) \oplus TT_{g_E}.$$

Let $h \in \Gamma(S^2M)$. Because of Lemma 2.4.1, we can write $h = \tilde{f} \cdot g_E + \delta^*\omega + \tilde{h}$ where $\tilde{f} \in C^\infty(M)$, $\omega \in \Omega^1(M)$ and $\tilde{h} \in TT$. By Obata's eigenvalue estimate, $\Delta - \mu$ is invertible for any Einstein manifold with constant $\mu \neq 0$. Thus,

$$\begin{aligned}h =&\tilde{f} \cdot g_E + \delta^*\omega + \tilde{h}\\ =&(\Delta - \mu)f \cdot g_E + \nabla^2 f + \delta^*(\omega - \nabla f) + \tilde{h}\\ =&C_{g_E}(f) + \delta^*(\omega - \nabla f) + \tilde{h},\end{aligned}$$

where $f = (\Delta - \mu)^{-1}\tilde{f}$. This shows the assertion. □

Theorem 5.4.3. *Let (M, g_E) be an Einstein manifold with constant -1 which is Einstein-Hilbert stable. Then there exists a $C^{2,\alpha}$-neighbourhood $\mathcal{U}$ of g_E such that $\mu_+(g) \leq \mu_+(g_E)$ for all $g \in \mathcal{U}$. Moreover, equality holds if and only if g is also an Einstein metric with constant -1.*

Proof. We first show that the second variation of $\mu_+(g_E)$ is nonpositive. By the second variational formula in Proposition 5.3.3, it suffices to show that the

Einstein operator is nonnegative on $\delta_{g_E}^{-1}(0)$. By Lemma 5.4.2, we have the L^2-orthogonal decomposition

$$\delta_{g_E}^{-1}(0) = C_{g_E}(C^\infty(M)) \oplus TT_{g_E},$$

where $C_{g_E}\colon C^\infty(M) \to \delta_{g_E}^{-1}(0)$ is defined as $C_{g_E}(f) = (\Delta f + f)g_E - \nabla^2 f$. On TT, Δ_E is nonnegative by assumption. Recall that $\Delta_L = \Delta_E - 2\cdot \mathrm{id}$ on an Einstein manifold with constant -1. We thus have

$$\begin{aligned}\Delta_E C_{g_E}(f) &= \Delta_L C_{g_E}(f) + 2C_{g_E}(f)\\ &= \Delta_L((\Delta f + f)g_E + \nabla^2 f) + 2C_{g_E}(f)\\ &= (\Delta(\Delta f) + (\Delta f))g_E + (\nabla^2 \Delta f) + 2C_{g_E}(f)\\ &= C_{g_E}(\Delta f + 2f).\end{aligned}$$

Here we used Lemma 2.4.5. We see that $C_{g_E}(f)$ is an eigentensor of Δ_E with eigenvalue λ if and only if f is an eigenfunction of the Laplace-Beltrami operator with eigenvalue $\lambda - 2$. Therefore, Δ_E is positive on $C_{g_E}(C^\infty(M))$.

Next, we show that $\mu_+(g) \leq \mu_+(g_E)$ on the affine slice $(g_E + \delta_{g_E}^{-1}(0)) \cap \mathcal{U}$ and equality holds if and only if $g \in \mathcal{P}$. Let $\delta_{g_E}^{-1}(0) = T_{g_E}\mathcal{P} \oplus N$ where N is the L^2-orthogonal complement of $T_{g_E}\mathcal{P}$ in $\delta_{g_E}^{-1}(0)$. By the above arguments, $\mu_+''(g_E)$ is negative definite on N. We consider the map

$$\begin{aligned}E\colon \mathcal{P} \times N &\to \mathcal{S}_{g_E},\\ (\bar{g}, h) &\mapsto \bar{g} + h.\end{aligned}$$

By the inverse function theorem for Banach manifolds, this is a local diffeomorphism around $(g_E, 0)$ if we temporarily enlarge the involved spaces to $C^{2,\alpha}$-spaces. However, each metric in $\mathcal{P}$ is smooth. This follows from the fact that $\ker(\Delta_E|_{TT}) = T_{g_E}\mathcal{P}$ is smooth by elliptic regularity and the arguments in [Has12, Proposition 3.6] and [Bes08, Theorem 12.49]. Therefore, each $C^{2,\alpha}$-metric in $\mathcal{S}_{g_E}$ near g_E is of the form $\bar{g} + h$ and is smooth if and only if h is smooth. By Taylor expansion,

$$\begin{aligned}\mu_+(\bar{g} + h) &= \mu_+(\bar{g}) + \frac{d}{dt}\Big|_{t=0} \mu_+(\bar{g} + th) + \frac{1}{2}\frac{d^2}{dt^2}\Big|_{t=0} \mu_+(\bar{g} + th) + R(\bar{g}, h),\\ R(\bar{g}, h) &= \int_0^1 \left(\frac{1}{2} - t + \frac{1}{2}t^2\right) \frac{d^3}{dt^3}\mu_+(\bar{g} + th)dt.\end{aligned}$$

Now, we claim that there exists a constant C_1 such that for all $\bar{g} \in \mathcal{P}$ and $h \in N$,

$$\frac{d^2}{dt^2}\Big|_{t=0} \mu_+(\bar{g} + th) \leq -C_1 \|h\|_{H^1}^2. \tag{5.16}$$

We define the projection map

$$pr\colon \mathcal{M} \times N \to \Gamma(S^2M)$$

which maps a tuple (g, h) to the projection of h onto the second factor of the splitting $\delta_g^*(\Omega^1(M)) \oplus \delta_g^{-1}(0)$. Let $\bar{h} = pr(\bar{g}, h)$. Recall that the second variation is only nonzero on the second factor. Therefore,

$$\begin{aligned}\frac{d^2}{dt^2}\Big|_{t=0} \mu_+(\bar{g} + th) &= -\frac{1}{4\mathrm{vol}(M, \bar{g})}\int_M \langle \Delta_E \bar{h}, \bar{h}\rangle\, dV_{\bar{g}}\\ &= -\frac{1}{4\mathrm{vol}(M, \bar{g})}\int_M \langle \Delta_E h, h\rangle\, dV + \frac{1}{4\mathrm{vol}(M, \bar{g})}\int_M \langle \Delta_E \delta^*\omega, \delta^*\omega\rangle\, dV,\end{aligned} \tag{5.17}$$

where ω is a 1-form satisfying $h = \delta^*\omega + \bar{h}$. We now deal with the first term of the right hand side. Since at $(\Delta_E)_{g_E}$ is positive definite on N, we have

$$\begin{aligned}((\Delta_E)_{g_E} h, h)_{L^2(g_E)} &= \epsilon \|\nabla h\|^2_{L^2(g_E)} + (1-\epsilon)(\nabla^*\nabla h - \frac{2}{1-\epsilon}\mathring{R}h, h)_{L^2(g_E)} \\ &\geq C_2 \|h\|^2_{H^1(g_E)},\end{aligned}$$

where $\epsilon > 0$ is sufficiently small. By Taylor expansion,

$$\begin{aligned}((\Delta_E)_{\bar{g}} h, h)_{L^2(\bar{g})} =& ((\Delta_E)_{g_E} h, h)_{L^2(g_E)} \\ &+ \int_0^1 \frac{d}{dt}((\Delta_E)_{g_E + t(\bar{g}-g_E)} h, h)_{L^2(g_E + t(\bar{g}-g_E))} dt.\end{aligned}$$

We have

$$\begin{aligned}\left.\frac{d}{dt}\right|_{t=0} ((\Delta_E)_{g+tk} h, h)_{L^2(g+tk)} =& (\Delta'_E h, h)_{L^2} - 2(\Delta_E h, k \circ h)_{L^2} \\ &+ \frac{1}{2}(\Delta_E h, (\mathrm{tr} k) h)_{L^2},\end{aligned}$$

and by Lemma A.5,

$$\Delta'_E h = \nabla^2 k * h + \nabla k * \nabla h + k * \nabla^2 h + R * k * h.$$

We arrive, using integration by parts, at an estimate of the form

$$\left.\frac{d}{dt}\right|_{t=0} ((\Delta_E)_{g+tk} h, h)_{L^2(g+tk)} \leq C_3 \|k\|_{C^2} \|h\|_{H^1}.$$

Thus,

$$((\Delta_E)_{\bar{g}} h, h)_{L^2} \geq C_2 \|h\|^2_{H^1(g_E)} - C_3 \|\bar{g} - g_E\|_{C^2} \|h\|_{H^1}.$$

Therefore, if we choose the $C^{2,\alpha}$-neighbourhood small enough, we have a uniform upper estimate on the first term of (5.17). The second term can be estimated from above by

$$\frac{1}{4\mathrm{vol}(M, \bar{g})} \int_M \langle \Delta_E \delta^*\omega, \delta^*\omega \rangle \, dV \leq C_4 \|\delta^*\omega\|_{H^1}.$$

By Lemma 5.4.4 below, we can choose, given any $\epsilon > 0$, the neighbourhood $\mathcal{U}$ so small that

$$\|\delta^*\omega\|_{H^1} < \epsilon \|h\|_{H^1},$$

and this implies inequality (5.16). By Proposition 5.3.5, we have the uniform estimate

$$|R(\bar{g}, h)| \leq C_3 \|h\|_{C^{2,\alpha}} \|h\|^2_{H^1}.$$

This yields

$$\mu_+(\bar{g} + h) \leq \mu_+(g) - \frac{C_1}{2} \|h\|^2_{H^1} + C_3 \|h\|_{C^{2,\alpha}} \|h\|^2_{H^1}. \tag{5.18}$$

Therefore, if we choose the $C^{2,\alpha}$-neighbourhood small enough, the negative term in (5.18) dominates and we obtain the desired inequality. Equality implies $h = 0$ which means that $g \in \mathcal{P}$. On $\mathcal{P}$, μ_+ is constantly equal to $\mu_+(g_E)$.

By the Ebin-Palais slice theorem, every $g \in \mathcal{U}$ can be written as $g = \varphi^* \tilde{g}$ for some $\varphi \in \mathrm{Diff}(M)$ and $\tilde{g} \in \mathcal{S}$. By diffeomorphism invariance,

$$\mu_+(g) = \mu_+(\tilde{g}) \leq \mu_+(g_E),$$

and equality holds if and only if $g = \varphi^* \tilde{g}$, where $\tilde{g} \in \mathcal{P}$. This implies that g is also an Einstein manifold with constant -1. □

Lemma 5.4.4. *Let g_E and $N, \omega, \mathcal{P}$ as above. Then for every $\epsilon > 0$, there exists a $C^{2,\alpha}$-neighbourhood $\mathcal{U}$ of g_E such that for all $g \in \mathcal{U} \cap \mathcal{P}$, $h \in N$,*

$$\|\delta^* \omega\|_{H^1(g)} \leq \epsilon \|h\|_{H^1(g)} .$$

Proof. Let $h = \delta_g^* \omega_g + pr(g,h)$ be the g-dependant decomposition of h according to the splitting $\Gamma(S^2 M) = \delta_g^*(\Omega^1(M)) \oplus \delta_g^{-1}(0)$. Then we have

$$\delta_g h = \delta_g \delta_g^* \omega_g .$$

The 1-form ω_g decomposes as $\omega_g = \nabla f_g + \bar{\omega}_g$, where $f_g \in C^\infty(M)$ and $\bar{\omega}_g \in \delta_g^{-1}(0)$. By straightforward calculation, one sees that

$$\delta_g \delta_g^* \omega_g = \delta_g \delta_g^* \nabla f_g + \delta_g \delta_g^* \bar{\omega}_g = (\nabla^* \nabla) \nabla f_g + \frac{1}{2} (\nabla^* \nabla - \mathrm{Ric}) \bar{\omega}_g .$$

This shows that $\delta_g \delta_g^*$ acts as an elliptic operator on both parts of the decomposition, since on Einstein manifolds, $\nabla^* \nabla$ and $\nabla^* \nabla - \mathrm{Ric} = \Delta_H$ preserve both subspaces. Since $\delta_g \delta_g^* \omega_g = 0$ implies $\delta_g^* \omega_g = 0$, we may choose ω_g to be orthogonal to the kernel of $\delta_g \delta_g^*$. Therefore, by elliptic regularity,

$$\begin{aligned} \|\delta_g^* \omega_g\|_{H^1}^2 \leq \|\omega_g\|_{H^2}^2 &\leq \|\nabla f_g\|_{H^2}^2 + \|\bar{\omega}_g\|_{H^2}^2 \\ &\leq C_1 \|\delta_g \delta_g^* \nabla f_g\|_{L^2}^2 + C_2 \|\delta_g \delta_g^* \bar{\omega}_g\|_{L^2}^2 \\ &\leq C_3 \|\delta_g \delta_g^* \omega_g\|_{L^2}^2 = C_3 \|\delta_g h\|_{L^2}^2 . \end{aligned}$$

In the last inequality, we used the fact that the decomposition

$$\Omega^1(M) = \nabla(C^\infty(M)) \oplus \delta_g^{-1}(0)$$

is L^2-orthogonal. We calculate

$$\begin{aligned} \left.\frac{d}{dt}\right|_{t=0} \|\delta_{g+tk} h\|_{L^2(g+tk)}^2 =& 2(\delta' h, \delta h)_{L^2} - \int_M k((\delta h)^\sharp, (\delta h)^\sharp) \, dV \\ &+ \frac{1}{2} \int_M |\delta h|^2 \mathrm{tr} k \, dV. \end{aligned}$$

By Lemma A.3,

$$\delta' h = \nabla k * h + k * \nabla h,$$

and thus, we get

$$\left.\frac{d}{dt}\right|_{t=0} \|\delta_{g+tk}h\|^2_{L^2(g+tk)} \leq C_4 \|h\|^2_{H^1(g)} \|k\|_{C^1(g)} .$$

Therefore by Taylor expansion,

$$\|\delta_{g_E+k}h\|^2_{L^2(g_E+k)} = \int_0^1 \frac{d}{dt} \|\delta_{g+tk}h\|^2_{L^2(g+tk)}\, dt \leq C_5 \|h\|^2_{H^1(g)} \|k\|_{C^{2,\alpha}(g)} .$$

Finally, if we choose the neighbourhood $\mathcal{U}$ so small that

$$\|g - g_E\|_{C^{2,\alpha}} \leq \epsilon (C_3 \cdot C_5)^{-1}$$

for all $g \in \mathcal{U} \cap \mathcal{P}$,

$$\left\|\delta_g^* \omega_g\right\|^2_{H^1} \leq C_3 \|\delta_g h\|^2_{L^2} \leq C_3 \cdot C_5 \|h\|^2_{H^1} \|g - g_E\|_{C^{2,\alpha}} \leq \epsilon \|h\|^2_{H^1} ,$$

which proves the lemma. □

5.4.2 A Lojasiewicz-Simon Inequality and Transversality

This subsection is devoted to the proof of two theorems which are essential ingredients in the proof of dynamical stability in the next section. Here we use the slice theorem and a certain 2-parameter expansion. The error terms can be controlled by the estimates we developed in Section 5.3.

Theorem 5.4.5 (Optimal Lojasiewicz-Simon Inequality for μ_+). *Let (M, g_E) be an Einstein manifold with constant -1. Then there exists a $C^{2,\alpha}$-neighbourhood $\mathcal{U}$ of g_E and a constant $C > 0$ such that*

$$|\mu_+(g) - \mu_+(g_E)|^{1/2} \leq C \left\|\mathrm{Ric} + g + \nabla^2 f_g\right\|_{L^2}$$

for all $g \in \mathcal{U}$.

Later on, this theorem will ensure that the Ricci flow converges exponentially as $t \to \infty$.

Theorem 5.4.6 (Transversality). *Let (M, g_E) be an Einstein manifold with constant -1. Then there exists a $C^{2,\alpha}$-neighbourhood $\mathcal{U}$ of g_E and a constant $C > 0$ such that*

$$\|\mathrm{Ric} + g\|_{L^2} \leq C \left\|\mathrm{Ric} + g + \nabla^2 f_g\right\|_{L^2}$$

for all $g \in \mathcal{U}$.

This theorem ensures that the Ricci flow does not move too excessively in gauge direction. We will conclude that the flow converges in the strict sense without pulling back by a family of diffeomorphisms.

Proof of Theorem 5.4.5 and Theorem 5.4.6. By diffeomorphism invariance, it suffices to prove both theorems on a slice in the space of metrics. As before, we work on the affine slice $\mathcal{S}_{g_E} = \mathcal{U} \cap (g_E + \delta_{g_E}^{-1}(0))$. As in the previous section, let

$$\mathcal{P} = \{g \in \mathcal{S}_{g_E} \mid \mathrm{Ric}_g = -g\} ,$$

and let N be the L^2-orthogonal complement of $T_{g_E}\mathcal{P}$ in $\delta_{g_E}^{-1}(0)$. Then every metric $g \in \mathcal{S}_{g_E}$ can be uniquely written as $g = \bar{g} + h$ with $\bar{g} \in \mathcal{P}$ and $h \in N$, provided that $\mathcal{U}$ is small enough. Taylor expansion yields the estimates

$$|\mu_+(\bar{g}+h) - \mu_+(\bar{g})| \leq C_1 \|h\|_{H^2}^2 ,$$
$$\|\mathrm{Ric}_{\bar{g}+h} + \bar{g} + h\|_{L^2}^2 \leq C_2 \|h\|_{H^2}^2 .$$

In order to prove the two theorems, it therefore suffices to show

$$\left\|\mathrm{Ric}_{\bar{g}+h} + \bar{g} + h + \nabla^2 f_{\bar{g}+h}\right\|_{L^2}^2 \geq C_3 \|h\|_{H^2}^2$$

for some appropriate constant C_3. To obtain this estimate, we need a few lemmas.

Lemma 5.4.7. *Let $F(s,t)$ be a C^2-function on $0 \leq s,t \leq 1$ with values in a Fréchet-space. Then*

$$F(1,1) = F(1,0) + \frac{d}{dt}|_{t=0}F(0,t) + \int_0^1 (1-t)\frac{d^2}{dt^2}F(0,t)dt + \int_0^1\int_0^1 \frac{d^2}{dsdt}F(s,t)dsdt.$$

Proof. See [Has12, Lemma 4.3]. □

Lemma 5.4.8. *Let $g = \bar{g} + h \in \mathcal{S}_{g_E}$ as above. Then we have the 2-parameter expansion*

$$\mathrm{Ric}_g + g + \nabla^2 f_g = \frac{1}{2}(\Delta_E)_{g_E} h + O_1 + O_2,$$

where

$$O_1 = \int_0^1 (1-t)\frac{d^2}{dt^2}(\mathrm{Ric}_{g_E+th} + g_E + th + \nabla^2 f_{g_E+th})dt,$$
$$O_2 = \int_0^1\int_0^1 \frac{d^2}{dsdt}(\mathrm{Ric}_{g_E+s(\bar{g}-g_E)+th} + g_E + s(\bar{g}-g_E) + th + \nabla^2 f_{g_E+s(\bar{g}-g_E)+th})dsdt.$$

Proof. We apply Lemma 5.4.7 to the map

$$F(s,t) = \mathrm{Ric}_{g_E+s(g-g_E)+th} + g_E + s(g-g_E) + th + \nabla^2 f_{g_E+s(g-g_E)+th}$$

and use Lemma 5.2.5 (iii). □

Lemma 5.4.9. *Let $g = \bar{g} + h \in \mathcal{S}_{g_E}$ as above. Then there exists a $C^{2,\alpha}$-neighbourhood and a constant $C > 0$ such that*

$$\|O_1\|_{L^2} \leq C \|h\|_{C^{2,\alpha}} \|h\|_{H^2} ,$$
$$\|O_2\|_{L^2} \leq C \|\bar{g} - g_E\|_{C^{2,\alpha}} \|h\|_{H^2} .$$

hold in this neighbourhood.

Proof. Let dot be t-derivatives and prime be s-derivatives. Put $k = g - g_E$. By Lemma A.3 and Lemma A.5, we have

$$
\begin{aligned}
(\mathrm{Ric}_g + g)^{\cdot\,\prime} &= \nabla^2 k * h + \nabla k * \nabla h + k * \nabla^2 h + R * k * h, \\
(\nabla f_g)^{\cdot\,\prime} &= \nabla k * h * \nabla f + k * \nabla h * \nabla f + \nabla f' * \nabla h + \nabla k * \nabla \dot{f} + \nabla^2 \dot{f}'.
\end{aligned}
$$

The estimate for O_2 follow from the Hölder inequality, Lemma 5.3.2 and Lemma 5.3.4. The other estimate is shown analogously. □

Let us now continue the main proof. By Lemma 5.4.8 and since $(\Delta_E)_{g_E}|_N$ is injective, we have

$$
\begin{aligned}
\left\|\mathrm{Ric}_{\bar{g}+h} + \bar{g} + h + \nabla^2 f_{\bar{g}+h}\right\|_{L^2}^2 \geq& \frac{1}{4}\left\|(\Delta_E)_{g_E} h\right\|_{L^2}^2 - \langle O_1 + O_2, (\Delta_E)_{g_E} h\rangle \\
\geq& C_4 \left\|h\right\|_{H^2}^2 - C_5(\left\|O_1\right\|_{L^2} + \left\|O_2\right\|_{L^2}) \left\|h\right\|_{H^2}.
\end{aligned}
$$

If we choose the neighbourhood small enough, Lemma 5.4.9 yields

$$
\left\|\mathrm{Ric}_{\bar{g}+h} + \bar{g} + h + \nabla^2 f_{\bar{g}+h}\right\|_{L^2}^2 \geq C_6 \left\|h\right\|_{H^2}^2,
$$

which finishes the proof of both theorems. □

5.4.3 Dynamical Stability and Instability

Lemma 5.4.10 (Estimates for $t \leq 1$). *Let (M, g_E) be an Einstein manifold with constant -1 and let $k \geq 2$. Then for all $\epsilon > 0$ there exists a $\delta > 0$ such that if $\left\|g_0 - g_E\right\|_{C^{k+2}_{g_E}} < \delta$, the Ricci flow starting at g_0 exists on $[0, 1]$ and satisfies*

$$
\left\|g(t) - g_E\right\|_{C^k_{g_E}} < \epsilon
$$

for all $t \in [0, 1]$.

Proof. The Riemann curvature tensor and the Ricci tensor evolve under the standard Ricci flow as $\partial_t R = -\Delta R + R * R$ and $\partial_t \mathrm{Ric} = -\Delta \mathrm{Ric} + R * \mathrm{Ric}$ (see e.g. [Bre10]). Under the flow $\dot{g}(t) = -2(\mathrm{Ric}_{g(t)} + g(t))$, they evolve as

$$
\partial_t R = -\Delta R + R * R - 4R, \qquad \partial_t \mathrm{Ric} = -\Delta \mathrm{Ric} + R * \mathrm{Ric}.
$$

The additional term $-4R$ comes from rescaling whereas the evolution equation for the Ricci tensor does not change because of scale-invariance. Therefore, we get the evolution inequalities

$$
\partial_t |\nabla^i R|^2 \leq -\Delta |\nabla^i R|^2 + \sum_{j=0}^{i} C_{ij} |\nabla^{i-j} R| |\nabla^j R| |\nabla^i R|,
$$

$$
\partial_t |\nabla^i (\mathrm{Ric} + g)|^2 \leq -\Delta |\nabla^i (\mathrm{Ric} + g)|^2 + \sum_{j=0}^{i} \tilde{C}_{ij} |\nabla^{i-j} R| |\nabla^j \mathrm{Ric}| |\nabla^i (\mathrm{Ric} + g)|.
$$

Here, all covariant derivatives, Laplacians and norms are taken with respect to $g(t)$. By the maximum principle for scalars (see e.g. [CCG$^+$08, Theorem 10.2]), there exists a $\tilde{K}(K,n,k) < \infty$ such that if $g(t)$ is a Ricci flow on $[0,T]$ with $T \leq 1$ and

$$\sup_{p\in M} |R_{g(t)}|_{g(t)} \leq K, \qquad \sup_{p\in M} |\nabla^i R_{g(0)}|_{g(0)} \leq K$$

for all $t \in [0,T]$ and $i \leq k$, then

$$\sup_{p\in M} |\nabla^i R_{g(t)}|_{g(t)} \leq \tilde{K}$$

for all $t \in [0,T]$ and $i \leq k$. Again by the maximum principle, there exists for each $\tilde{\epsilon} > 0$ a $\tilde{\delta}(\tilde{\epsilon}, \tilde{K}, n, k) > 0$ such that, if

$$\sup_{p\in M} |\nabla^i(\mathrm{Ric}_{g(0)} + g(0))|_{g(0)} \leq \tilde{\delta}$$

for all $i \leq k$, we have

$$\sup_{p\in M} |\nabla^i(\mathrm{Ric}_{g(t)} + g(t))|_{g(t)} \leq \tilde{\epsilon}$$

for all $t \in [0,T]$ and $i \leq k$. If we choose the ϵ-neighbourhood small enough such that the C^k-norms with respect to g_E and $g(t)$ differ at most by a factor 2,

$$\frac{d}{dt}\|g(t) - g_E\|_{C^k_{g_E}} \leq \left\|2(\mathrm{Ric}_{g(t)} + g(t))\right\|_{C^k_{g_E}} \leq 4\left\|(\mathrm{Ric}_{g(t)} + g(t))\right\|_{C^k_{g(t)}}.$$

Let $\bar{\delta} > 0$ be small enough and let

$$K = \sup\left\{\|R_g\|_{C^0_g} \mid \|g - g_E\|_{C^k_{g_E}} < \epsilon\right\} + \sup\left\{\|R_g\|_{C^k_g} \mid \|g - g_E\|_{C^{k+2}_{g_E}} < \bar{\delta}\right\}.$$

Let $\tilde{K} = \tilde{K}(K,n,k)$, $\tilde{\delta} = \tilde{\delta}(\tilde{K}, \tilde{\epsilon}, n, k)$ and $\delta_1 < \bar{\delta}$ be so small that for

$$\|g_0 - g_E\|_{C^{k+2}_{g_E}} \leq \delta_1,$$

we have

$$\sup_{p\in M, i\leq k} |\nabla^i(\mathrm{Ric}_{g_0} + g_0)|_{g(0)} \leq \tilde{\delta}, \qquad \|g_0 - g_E\|_{C^k_{g_E}} \leq \frac{\epsilon}{4},$$

and the Ricci flow starting at g_0 satisfies

$$\sup_{(p,t)\in M\times[0,T], i\leq k} |\nabla^i(\mathrm{Ric}_{g(t)} + g(t))|_{g(t)} \leq \tilde{\epsilon} = \frac{\epsilon}{16(k+1)}.$$

Let $T \in [0,\infty]$ be the maximal interval such that the Ricci flow starting at g_0 exists on $[0,T)$ and satisfies

$$\|g(t) - g_E\|_{C^k_{g_E}} < \epsilon$$

for all $t \in [0,T)$. Suppose that $T \leq 1$. Then

$$\begin{aligned}
\|g(t) - g_E\|_{C^k_{g_E}} &\leq \|g_0 - g_E\|_{C^k_{g_E}} + \int_0^T \frac{d}{dt}\|g(t) - g_E\|_{C^k_{g_E}}\, dt \\
&\leq \|g_0 - g_E\|_{C^k_{g_E}} + 4\int_0^T \left\|(\mathrm{Ric}_{g(t)} + g(t))\right\|_{C^k_{g(t)}} dt \\
&\leq \delta + 4(k+1)\tilde{\epsilon} \leq \frac{\epsilon}{2},
\end{aligned}$$

which contradicts the maximality of T. □

Lemma 5.4.11. *Let $g(t)$, $t \in [0,T]$ be a solution of the Ricci flow (5.5) and suppose that*

$$\sup_{p\in M} |R_{g(t)}|_{g(t)} \leq T^{-1} \qquad \forall t \in [0,T].$$

Then for each $k \geq 1$, there exists a constant $C(k)$ such that

$$\sup_{p\in M} |\nabla^k R_{g(t)}|_{g(t)} \leq C(k) \cdot T^{-1} t^{-k/2} \qquad \forall t \in (0,T].$$

Proof. This is a well-known result for the standard Ricci flow. For the sake of completeness, we redo the proof of Hamilton in [Ham95, Theorem 7.1] which also works for the flow (5.5). From the evolution equation $\partial_t R = -\Delta R + R * R - 4R$, we obtain the evolution inequality

$$\partial_t |\nabla^k R|^2 \leq -\Delta |\nabla^k R|^2 - 2|\nabla^{k+1} R|^2 + \sum_{j=0}^{k} C_{jk} |\nabla^j R||\nabla^{k-j} R||\nabla^k R|.$$

We will now use the $|\nabla^{k+1}R|^2$-term we omitted in the proof of the previous lemma. In particular, we have

$$\partial_t |R|^2 \leq -\Delta |R|^2 - 2|\nabla R|^2 + C_{00}|R|^3,$$
$$\partial_t |\nabla R|^2 \leq -\Delta |\nabla R|^2 - 2|\nabla^2 R|^2 + 2C_{01}|R||\nabla R|^2.$$

Let now F be the function

$$F = t|\nabla R|^2 + A|R|^2,$$

where A is some large constant. We have

$$\partial_t F \leq -\Delta F + (C_1 t|R| - 2A)|\nabla R|^2 + C_2 A |R|^3.$$

By assumption, $|R| \leq T^{-1}$ and $tT \leq 1$. Thus, if we take $2A \geq C_1$,

$$\partial_t F \leq -\Delta F + C_3 T^{-3}.$$

Since $F(0) \leq C_4 T^{-2}$, the maximum principle yields

$$F \leq C_4 T^{-2} + t C_3 T^{-3} \leq C_5 T^{-2}$$

for $t \leq T$. By definition of F, we thus have

$$t|\nabla R|^2 \leq F \leq C_5 T^{-2}.$$

This shows the assertion for $k = 1$. Now we proceed by induction. Suppose that the lemma holds for a fixed $k \in \mathbb{N}$. Then by the evolution inequalities and induction hypothesis,

$$\begin{aligned} \partial_t |\nabla^k R|^2 \leq & -\Delta |\nabla^k R|^2 - 2|\nabla^{k+1} R|^2 + C_6 T^{-3} t^{-k}, \\ \partial_t |\nabla^{k+1} R|^2 \leq & -\Delta |\nabla^{k+1} R|^2 - 2|\nabla^{k+1} R|^2 + C_7 T^{-1} |\nabla^{k+1} R|^2 \\ & + C_8 T^{-2} |\nabla^{k+1} R| t^{-(k+1)/2}. \end{aligned}$$

From

$$T^{-2}|\nabla^{k+1}R|t^{-(k+1)/2} \leq \frac{1}{2}(T^{-1}|\nabla^{k+1}R|^2 + T^{-3}t^{k+1}),$$

we obtain

$$\partial_t|\nabla^{k+1}R|^2 \leq -\Delta|\nabla^{k+1}R|^2 + C_9T^{-1}|\nabla^{k+1}R|^2 + C_{10}T^{-3}t^{k+1}.$$

Now we define

$$F_k = t|\nabla^{k+1}|^2 + A_k|\nabla^k R|^2,$$

where A_k is some large constant. Then

$$\partial_t F_k \leq -\Delta F_k + (C_{11}tT^{-1} - 2A_k)|\nabla^{k+1}R|^2 + C_{12}A_kT^{-3}t^{-k},$$

and if $t \leq T$ and $A_k \geq 2C_{11}$,

$$\partial_t F_k \leq \Delta F_k + C_{13}T^{-3}t^{-k}.$$

Since $F_k(0) \leq C_{14}T^{-2}t^{-k}$, the maximum principle yields

$$F_k \leq C_{14}T^{-2}t^{-k} + C_{13}T^{-3}t^{-k+1} \leq C_{15}T^{-2}t^{-k},$$

where we used $t \leq T$. Thus,

$$t|\nabla^{k+1}R|^2 \leq F_k \leq C_{15}T^{-2}t^{-k},$$

and the induction step is completed. This proves the lemma. □

Remark 5.4.12. For a given Ricci flow $g(t)$, the previous lemma and a bootstrap argument imply the following: Suppose we have a uniform bound

$$\sup_{p\in M}|R_{g(t)}|_{g(t)} \leq K \qquad \forall t \in [0,T].$$

Then for each $k \geq 1$ and $\delta > 0$, there exists a constant $C(k,\delta)$ such that

$$\sup_{p\in M}|\nabla^k R_{g(t)}|_{g(t)} \leq C(k,\delta)\cdot K \qquad \forall t \in [\delta,T].$$

Theorem 5.4.13 (Dynamical stability). *Let (M, g_E) be a compact Einstein manifold with constant -1 which is Einstein-Hilbert stable and satisfies the integrability condition. Let $k \geq 3$. Then for every C^k-neighbourhood $\mathcal{U}$ of g_E, there exists a C^{k+2}-neighbourhood $\mathcal{V} \subset \mathcal{U}$ of g_E such that the following holds: For any metric $g_0 \in \mathcal{V}$ the Ricci flow* (5.5) *starting at g_0 stays in $\mathcal{U}$ for all time. Moreover, the Ricci flow converges to some Einstein metric $g_\infty \in \mathcal{U}$ with constant -1 as $t \to \infty$. The convergence is exponentially, i.e. there exist constants $C_1, C_2 > 0$ such that for all $t \geq 0$,*

$$\|g(t) - g_\infty\|_{C^k_{g_E}} \leq C_1 e^{-C_2 t}.$$

Proof. We write $\mathcal{B}^k_\epsilon$ for the ϵ-ball around g_E with respect to the $C^k_{g_E}$-norm. Without loss of generality, we may assume that $\mathcal{U} = \mathcal{B}^k_\epsilon$ and $\epsilon > 0$ is so small that Theorems 5.4.3, 5.4.5 and 5.4.6 are satisfied. By Lemma 5.4.10, we can choose $\mathcal{V}$ so small that any Ricci flow starting in $\mathcal{V}$ exists on $[0,1]$ and stays in $\mathcal{B}^k_{\epsilon/4}$ up to time 1.

Let now $g_0 \in \mathcal{V}$ be arbitrary and $T \in (1,\infty)$ be the maximal time such that the Ricci flow starting at g_0 exists on $[0,T)$ and stays in $\mathcal{U}$ for all $t \in [0,T)$. Then

$$\begin{aligned}
\|g(t) - g_E\|_{C^k_{g_E}} &\leq \|g(1) - g_E\|_{C^k_{g_E}} + \int_1^T \frac{d}{dt}\|g(t) - g_E\|_{C^k_{g_E}}\, dt \\
&\leq \frac{\epsilon}{4} + \int_1^T \left\|2(\mathrm{Ric}_{g(t)} + g(t))\right\|_{C^k_{g_E}} dt \\
&\leq \frac{\epsilon}{4} + 4\int_1^T \left\|\mathrm{Ric}_{g(t)} + g(t)\right\|_{C^k_{g(t)}} dt.
\end{aligned}$$

Here we assumed that ϵ is so small that the C^k-norms defined by $g(t)$ and g_E differ at most by a factor 2. By Remark 5.4.12, we have uniform bounds

$$\sup_{p\in M} |\nabla^i R_{g(t)}|_{g(t)} \leq C(i) \qquad \forall t \in [1,T).$$

By interpolation inequalites for tensors (c.f. [Ham82, Corollary 12.7]), we therefore get, if we fix some $\beta \in (0,1)$,

$$\left\|\mathrm{Ric}_{g(t)} + g(t)\right\|_{C^k_{g(t)}} \leq C_S \left\|\mathrm{Ric}_{g(t)} + g(t)\right\|_{H^l_{g(t)}} \leq C_S C_1 \left\|\mathrm{Ric}_{g(t)} + g(t)\right\|^\beta_{L^2_{g(t)}}.$$

Here $l \geq k$ is some sufficiently large number and C_S is the constant from Sobolev embedding. Suppose that $g(t)$ is not an Einstein metric with constant -1 (otherwise the flow is trivial). Then by Theorems 5.4.3, 5.4.5 and 5.4.6 and the first variation of μ_+,

$$\begin{aligned}
-\frac{d}{dt}|\mu_+(g(t)) - \mu_+(g_E)|^{\beta/2} &= \frac{\beta}{2}|\mu_+(g(t)) - \mu_+(g_E)|^{\beta/2-1}\frac{d}{dt}\mu_+(g(t)) \\
&= \frac{\beta}{2}|\mu_+(g(t)) - \mu_+(g_E)|^{\beta/2-1}\int_M |\mathrm{Ric}_{g(t)} + g(t) + \nabla^2 f_{g(t)}|^2 e^{-f_{g(t)}}\, dV_{g(t)} \\
&\geq C_2|\mu_+(g(t)) - \mu_+(g_E)|^{\beta/2-1}\left\|\mathrm{Ric}_{g(t)} + g(t) + \nabla^2 f_{g(t)}\right\|^2_{L^2_{g(t)}} \\
&\geq C_3 \left\|\mathrm{Ric} + g\right\|^\beta_{L^2_{g(t)}}.
\end{aligned}$$

Hence by integration and monotonicity of μ_+ along the flow,

$$\begin{aligned}
\int_1^T \left\|\mathrm{Ric}_{g(t)} + g(t)\right\|_{C^k_{g(t)}} &\leq C_4 \int_1^T \left\|\mathrm{Ric}_{g(t)} + g(t)\right\|^\beta_{L^2_{g(t)}} dt \\
&\leq C_5|\mu_+(g(1)) - \mu_+(g_E)|^{\beta/2} \\
&\leq C_5|\mu_+(g_0) - \mu_+(g_E)|^{\beta/2} \leq \frac{\epsilon}{16},
\end{aligned} \tag{5.19}$$

provided we have chosen $\mathcal{V}$ small enough. We thus obtain

$$\|g(t) - g_E\|_{C^k_{g_E}} \leq \frac{\epsilon}{2},$$

which contradicts the maximality of T. Therefore, $T = \infty$ and for all $t \geq 0$, we have

$$\|g(t) - g_E\|_{C^k_{g_E}} < \epsilon,$$
$$\int_0^\infty \|\dot{g}(t)\|_{C^k_{g_E}} \, dt < \infty.$$

It follows that $g(t) \to g_\infty$ as $t \to \infty$. Along the flow, we have, by Theorem 5.4.5, $-\frac{d}{dt}|\mu_+(g(t)) - \mu_+(g_E)| \geq C_6|\mu_+(g(t)) - \mu_+(g_E)|$. Thus,

$$|\mu_+(g(t_2)) - \mu_+(g_E)| \leq e^{-C_6(t_2-t_1)}|\mu_+(g(t_1)) - \mu_+(g_E)|,$$

which shows that $\mu_+(g_\infty) = \mu_+(g_E)$ and by Theorem 5.4.3, $\mathrm{Ric}_{g_\infty} = -g_\infty$. The convergence is exponential since for $t_1 < t_2$,

$$\begin{aligned}\|g(t_1) - g(t_2)\|_{C^k_{g_E}} \leq & C_7|\mu_+(g(t_1)) - \mu_+(g_E)|^{\beta/2} \\ \leq & C_7 e^{-\frac{\beta}{2}C_6 t_1}|\mu_+(g_0) - \mu_+(g_E)|,\end{aligned}$$

where the first inequality follows from arguments as in (5.19). The assertion follows from $t_2 \to \infty$. □

Theorem 5.4.14 (Dynamical instability). *Let (M, g_E) be an Einstein manifold with constant -1 which satisfies the integrability condition. If (M, g_E) is Einstein-Hilbert unstable, there exists a nontrivial ancient Ricci flow emerging from it, i.e. there is a Ricci flow $g(t)$, $t \in (-\infty, 0]$ such that $\lim_{t\to-\infty} g(t) = g_E$.*

Proof. Since (M, g_E) is Einstein-Hilbert unstable, it cannot be a local maximum of μ_+. Let $g_i \to g_E$ in C^k and suppose that $\mu_+(g_i) > \mu_+(g_E)$ for all i. Let $g_i(t)$ be the Ricci flow (5.5) starting at g_i. Then by Lemma 5.4.10, $\bar{g}_i = g_i(1)$ converges to g_E in C^{k-2} and by monotonicity, $\mu_+(\bar{g}_i) > \mu_+(g_E)$ as well. Let $\epsilon > 0$ be so small that Theorems 5.4.5 and 5.4.6 both hold on $\mathcal{B}^{k-2}_{2\epsilon}$. Theorem 5.4.5 yields the differential inequality

$$\frac{d}{dt}(\mu_+(g_i(t)) - \mu_+(g_E)) \geq C_1(\mu_+(g_i(t)) - \mu_+(g_E)),$$

from which we obtain

$$(\mu_+(g_i(t)) - \mu_+(g_E))e^{C_1(s-t)} \leq (\mu_+(g_i(s)) - \mu_+(g_E)), \tag{5.20}$$

as long as g_i stays in $\mathcal{B}^{k-2}_{2\epsilon}$. Thus, there exists a time t_i such that

$$\|g_i(t_i) - g_E\|_{C^{k-2}} = \epsilon.$$

Now observe that Lemma 5.4.10 holds if we replace 1 by any other time. Thus, $t_i \to \infty$ because $g_i(t_i) \to g_E$ in C^{k-2} if t_i was bounded. By interpolation,

$$\left\|\mathrm{Ric}_{g_i(t)} - g_i(t)\right\|_{C^{k-2}} \leq C_2 \left\|\mathrm{Ric}_{g_i(t)} - g_i(t)\right\|^\beta_{L^2} \tag{5.21}$$

for some $\beta \in (0,1)$. By Theorems 5.4.5 and 5.4.6, we have the differential inequality

$$\frac{d}{dt}(\mu_+(g_i(t)) - \mu_+(g_E))^{\beta/2} \geq C_3 \left\|\mathrm{Ric}_{g_i(t)} + g_i(t)\right\|^\beta_{L^2}, \tag{5.22}$$

if $\mu_+(g_i(t)) > \mu_+(g_E)$. Therefore, by the triangle inequality and by integration,

$$\epsilon = \|g_i(t_i) - g_E\|_{C^{k-2}} \leq \|\bar{g}_i - g_E\|_{C^{k-2}} + C_4(\mu_+(g_i(t_i)) - \mu_+(g_E))^{\beta/2}. \quad (5.23)$$

Now put $g_i^s(t) := g_i(t + t_i)$, $t \in [T_i, 0]$, where $T_i = 1 - t_i \to -\infty$. We have

$$\|g_i^s(t) - g_E\|_{C^{k-2}} \leq \epsilon \qquad \forall t \in [T_i, 0],$$
$$g_i^s(T_i) \to g_E \text{ in } C^{k-2}.$$

Because the embedding $C^{k-3}(M) \subset C^{k-2}(M)$ is compact, we can choose a subsequence of the g_i^s, converging in $C^{k-3}_{loc}(M \times (-\infty, 0])$ to an ancient Ricci flow $g(t)$, $t \in (-\infty, 0]$. From taking the limit $i \to \infty$ in (5.23), we have $\epsilon \leq C_4(\mu_+(g(0)) - \mu_+(g_E))^{\beta/2}$ which shows that the Ricci flow is nontrivial. For $T_i \leq t$, we have, by (5.21) and (5.22),

$$\begin{aligned}\|g_i^s(T_i) - g_i^s(t)\|_{C^{k-3}} \leq & C_5(\mu_+(g_i(t + t_i)) - \mu_+(g_E))^{\beta/2} \\ \leq & C_5(\mu_+(g_i(t_i)) - \mu_+(g_E))^{\beta/2} e^{C_1 t} = C_6 e^{C_1 t}.\end{aligned}$$

Thus,

$$\|g_E - g(t)\|_{C^{k-3}} \leq \|g_E - g_i^s(T_i)\|_{C^{k-3}} + C_6 e^{C_1 t} + \|g_i^s(t) - g(t)\|_{C^{k-3}}.$$

It follows that $\|g_E - g(t)\|_{C^{k-3}} \to 0$ as $t \to -\infty$. □

Remark 5.4.15. All known compact negative Einstein manifolds are Einstein-Hilbert stable (see e.g. [Dai07]). Moreover, no nonintegrable infinitesimal Einstein deformations are known in the negative case. Therefore, all known examples are dynamically stable by Theorem 5.4.13.

It would be very interesting to generalize these theorems to the noncompact case. Stability of the hyperbolic space under Ricci flow was studied in [SSS11; Bam11]. The more general case of symmetric spaces of noncompact type was studied in [Bam10]. There, the nonnegativity of the Einstein operator plays an important role.

5.5 The Nonintegrable Case

The integrability condition we assumed is a strong condition and one cannot expect that it holds in general. Luckily we were able to get rid of this condition. In this section, we prove dynamical stability and instability theorems without the integrabilty assumption. In contrast to the previous results, we obtain convergence modulo diffeomorphism and the convergence rate is polynomially.

Recall that the integrability condition was nessecary in proving Theorems 5.4.3, 5.4.5 and 5.4.6. In this section we prove analogoues of Theorems 5.4.3 and 5.4.5.

5.5.1 Local Maximum of the Expander Entropy

Here we give a different characterization of local maximality of μ_+. We use the local decomposition of the space of metrics stated in Theorem 2.6.1 and the observation that the μ_+-functional can be explicitly evaluated on metrics of constant scalar curvature.

Theorem 5.5.1. *Let (M, g_E) be a compact Einstein manifold with constant -1. Then g_E is a maximum of the μ_+-functional in a $C^{2,\alpha}$-neighbourhood if and only if g is a local maximum of the Yamabe functional in a $C^{2,\alpha}$-neighbourhood. In this case, any metric sufficiently close to g_E with $\mu_+(g) = \mu_+(g_E)$ is Einstein with constant -1.*

Proof. Let $c = \mathrm{vol}(M, g_E)$ and write

$$\mathcal{C} = \{g \in \mathcal{M} | \mathrm{scal}_g \text{ is constant}\},$$
$$\mathcal{C}_c = \{g \in \mathcal{M} | \mathrm{scal}_g \text{ is constant and } \mathrm{vol}(M, g) = c\}.$$

Since $\frac{\mathrm{scal}_{g_E}}{n-1} \notin \mathrm{spec}_+(\Delta_{g_E})$, Theorem 2.6.1 asserts that the map

$$\Phi \colon C^\infty(M) \times \mathcal{C}_c \to \mathcal{M}$$
$$(v, g) \mapsto v \cdot g$$

is a local ILH-diffeomorphism around $(1, g_E)$. Recall also that any metric $g \in \mathcal{C}$ sufficiently close to g_E is a Yamabe metric.

By the proof of Lemma 5.2.5 (i), the minimizer $f_{\bar{g}}$ realizing $\mu_+(\bar{g})$ is constant and equal to $\log(\mathrm{vol}(M, \bar{g}))$ if $\bar{g} \in \mathcal{C}$. Thus, $\mu_+(\bar{g}) = \frac{1}{2}\mathrm{scal}_{\bar{g}} - \log(\mathrm{vol}(M, \bar{g}))$. If g_E is not a local maximum of the Yamabe functional, there exist metrics $g_i \in \mathcal{C}_c$, $g_i \to g_E$ in $C^{2,\alpha}$ which have the same volume but larger scalar curvature than g_E. Thus, also $\mu_+(g_i) > \mu_+(g_E)$ which causes the contradiction.

If g_E is a local maximum of the Yamabe functional, it is a local maximum of μ_+ restricted to $\mathcal{C}_c$. Any other metric $\bar{g} \in \mathcal{C}_c$ satisfying $\mu_+(\bar{g}) = \mu_+(g_E)$ is also a local maximum of the Yamabe functional. In particular, $\bar{g}$ is a critical point of the total scalar curvature restricted to $\mathcal{C}_c$ and the scalar curvature is equal to $-n$. By Proposition 2.6.2, $\bar{g}$ is an Einstein manifold with constant -1. For $\alpha \cdot \bar{g}$, where $\alpha > 0$ and $\bar{g} \in \mathcal{C}_c$ sufficiently close to g_E, we have

$$\begin{aligned}
\mu_+(\alpha \cdot \bar{g}) &= \frac{1}{2}\mathrm{scal}_{\alpha \cdot \bar{g}} - \log(\mathrm{vol}(M, \alpha \cdot \bar{g})) \\
&= \frac{1}{2\alpha}\mathrm{scal}_{\bar{g}} - \log(\alpha^{n/2}\mathrm{vol}(M, \bar{g})) \\
&\leq \frac{1}{2\alpha}\mathrm{scal}_{g_E} - \frac{n}{2}\log(\alpha) - \log(\mathrm{vol}(M, \bar{g})) \\
&= -\frac{n}{2}\left(\frac{1}{\alpha} + \log(\alpha)\right) - \log(\mathrm{vol}(M, g_E)) \\
&\leq -\frac{n}{2} - \log(\mathrm{vol}(M, g_E)) = \mu_+(g_E),
\end{aligned}$$

which shows that g_E is also a local maximum of μ_+ restricted to $\mathcal{C}$ and equality occurs if and only if $\alpha = 1$ and $\mu_+(\bar{g}) = \mu_+(g_E)$.

It remains to investigate the variation of μ_+ in the direction of volume-preserving conformal deformations. Let $h = v \cdot \bar{g}$, where $\bar{g} \in \mathcal{C}$ and $v \in C^\infty(M)$ with $\int_M v \, dV_{\bar{g}} = 0$. Then

$$\begin{aligned}
\frac{d}{dt}\bigg|_{t=0} \mu_+(\bar{g} + th) &= -\frac{1}{2}\int_M \langle \mathrm{Ric}_{\bar{g}} + \bar{g}, h\rangle e^{-f_{\bar{g}}} \, dV \\
&= -\frac{1}{2}\fint_M (\mathrm{scal}_{\bar{g}} + n)v \, dV = 0,
\end{aligned}$$

since $f_{\bar{g}}$ is constant. The second variation equals

$$\begin{aligned}
&\left.\frac{d^2}{dt^2}\right|_{t=0} \mu_+(\bar{g}+th)\\
&= -\left.\frac{1}{2}\frac{d}{dt}\right|_{t=0} \int_M \langle \mathrm{Ric}_{\bar{g}+th} + \bar{g} + th + \nabla^2 f_{\bar{g}+th}, h\rangle_{\bar{g}+th} e^{-f_{\bar{g}+th}}\, dV_{\bar{g}+th}\\
&= -\frac{1}{2}\fint_M \left\langle \left.\frac{d}{dt}\right|_{t=0} (\mathrm{Ric}_{\bar{g}+th} + \bar{g} + th + \nabla^2 f_{\bar{g}+th}), h\right\rangle_{\bar{g}} dV_{\bar{g}}\\
&\quad + \fint_M \langle \mathrm{Ric}_{\bar{g}} + \bar{g}, h\circ h\rangle_{\bar{g}}\, dV_{\bar{g}} - \frac{1}{2}\fint_M \langle \mathrm{Ric}_{\bar{g}} + \bar{g}, h\rangle \left(-f' + \frac{1}{2}\mathrm{tr} h\right) dV_{\bar{g}}.
\end{aligned}$$

By the first variation of the Ricci tensor,

$$\begin{aligned}
-&\frac{1}{2}\fint_M \langle \mathrm{Ric}' + h, h\rangle\, dV_{\bar{g}}\\
&= -\frac{1}{2}\fint_M \left\langle \frac{1}{2}\Delta_L h - \delta^*(\delta h) - \frac{1}{2}\nabla^2 \mathrm{tr} h, h\right\rangle dV_{\bar{g}} - \frac{n}{2}\fint_M v^2\, dV_{\bar{g}}\\
&= -\frac{1}{2}\fint_M \left\langle (\Delta v)\bar{g} + \left(1-\frac{n}{2}\right)\nabla^2 v, v\cdot \bar{g}\right\rangle dV_{\bar{g}} - \frac{n}{2}\fint_M v^2\, dV_{\bar{g}}\\
&= -\frac{n-1}{2}\fint_M |\nabla v|^2\, dV_{\bar{g}} - \frac{n}{2}\fint_M v^2\, dV_{\bar{g}}.
\end{aligned}$$

By differentiating Euler-Lagrange equation (5.7), we have

$$(\Delta + 1)f' = \frac{1}{2}((n-1)\Delta v - \mathrm{scal}_{\bar{g}} v). \tag{5.24}$$

Thus,

$$\begin{aligned}
-\frac{1}{2}\int_M \left\langle \left.\frac{d}{dt}\right|_{t=0} \nabla^2 f_{\bar{g}+th}, h\right\rangle e^{-f_{\bar{g}}}\, dV &= \frac{1}{2}\fint_M \Delta f'\cdot v\, dV\\
&= \frac{1}{2}\fint_M (\Delta+1)f'\cdot v\, dV - \frac{1}{2}\fint_M f'\cdot v\, dV\\
&= \frac{1}{4}\fint_M [(n-1)\Delta v - \mathrm{scal}_{\bar{g}} v]v\, dV - \frac{1}{2}\fint_M f'\cdot v\, dV.
\end{aligned}$$

Adding up, we obtain

$$\begin{aligned}
-&\frac{1}{2}\int_M \left\langle \left.\frac{d}{dt}\right|_{t=0} (\mathrm{Ric}_{\bar{g}+th} + \bar{g} + th + \nabla^2 f_{\bar{g}+th}), h\right\rangle_{\bar{g}} e^{-f_{\bar{g}}}\, dV_{\bar{g}}\\
&= -\frac{1}{4}\fint_M |\nabla v|^2\, dV_{\bar{g}} - \frac{1}{2}\left(n + \frac{\mathrm{scal}_{\bar{g}}}{2}\right)\fint_M v^2\, dV - \frac{1}{2}\fint_M f'\cdot v\, dV\\
&\le -C_1 \|v\|_{H^1}^2,
\end{aligned}$$

and this estimate is uniform in a small $C^{2,\alpha}$-neighbourhood of g_E. Here we have used that by (5.24), the L^2-scalar product of f' and v is positive. Given any $\epsilon > 0$, the remaining terms of the second variation can be estimated by

$$\int_M \langle \mathrm{Ric}_{\bar{g}} + \bar{g}, h\circ h\rangle_{\bar{g}} e^{-f_{\bar{g}}}\, dV_{\bar{g}} = (\mathrm{scal}_{\bar{g}} + n)\fint_M v^2\, dV \le \epsilon \|v\|_{L^2}^2$$

and

$$-\frac{1}{2}\int_M \langle \mathrm{Ric}_{\bar{g}} + \bar{g}, h\rangle \left(-f'_{\bar{g}} + \frac{1}{2}\mathrm{tr}h\right) e^{-f_{\bar{g}}}\, dV_{\bar{g}}$$
$$= -\frac{\mathrm{scal}_{\bar{g}} + n}{2} \fint_M v\left(-f'_{\bar{g}} + \frac{n}{2}v\right)\, dV \leq \epsilon \left\|v\right\|_{L^2}^2,$$

provided that the neighbourhood is small enough. In the last inequality, we used $\|f'\|_{L^2} \leq C_2 \|v\|_{L^2}$ which holds because of (5.24) and elliptic regularity. Thus, we have a uniform estimate

$$\left.\frac{d^2}{dt^2}\right|_{t=0} \mu_+(\bar{g} + tv\bar{g}) \leq -C_3 \left\|v\right\|_{H^1}^2.$$

Let now g be an arbitrary metric in a small $C^{2,\alpha}$-neighbourhood of g_E. By the above, it can be written as $g = \tilde{v}\cdot\tilde{g}$, where $(\tilde{v},\tilde{g}) \in C^\infty(M) \times \mathcal{C}_{g_E}$ is close to $(1, g_E)$. By substituting

$$v = \frac{\tilde{v} - \fint \tilde{v}\, dV_{\tilde{g}}}{\fint \tilde{v}\, dV_{\tilde{g}}}, \qquad \bar{g} = \left(\fint \tilde{v}\, dV_{\tilde{g}}\right)\tilde{g},$$

we can write $g = (1+v)\bar{g}$, where $\bar{g} \in \mathcal{C}$ is close to g_E and $v \in C^\infty_{\bar{g}}(M)$ is close to 0. Thus by Taylor expansion and Proposition 5.3.5,

$$\mu_+(g) = \mu_+(\bar{g}) + \frac{1}{2}\left.\frac{d^2}{dt^2}\right|_{t=0} \mu_+(\bar{g}+tv\bar{g}) + \int_0^1 \left(\frac{1}{2} - t + \frac{1}{2}t^2\right)\frac{d^3}{dt^3}\mu_+(\bar{g}+tv\bar{g})dt$$
$$\leq \mu_+(g_E) - C_4 \left\|v\right\|_{H^1}^2 + C_5 \left\|v\right\|_{C^{2,\alpha}} \left\|v\right\|_{H^1}^2.$$

Now if we choose the $C^{2,\alpha}$-neighbourhood small enough, $\mu_+(g) \leq \mu_+(g_E)$ and equality holds if and only if $v \equiv 0$ and $\mu_+(g) = \mu_+(g_E)$. As discussed earlier in the proof, this implies that g is Einstein with constant -1. □

5.5.2 A Lojasiewicz-Simon Inequality

For proving a gradient inequality in the nonintegrable case, we need to know that μ_+ is analytic. To show this, we use the implicit function theorem for Banach manifolds in the analytic category mentioned in [Koi83, Section 13]. Such arguments were also used in [SW13] for a result similar to the below lemma.

Lemma 5.5.2. *There exists a $C^{2,\alpha}$-neighbourhood $\mathcal{U}$ of g_E such that the map $g \mapsto \mu_+(g)$ is analytic on $\mathcal{U}$.*

Proof. Let $H(g,f) = -\Delta_g f - \frac{1}{2}|\nabla f|^2 + \frac{1}{2}\mathrm{scal}_g - f$ and consider the map

$$L\colon \mathcal{M}^{C^{2,\alpha}} \times C^{2,\alpha}(M) \to C^{0,\alpha}_{g_E}(M) \times \mathbb{R}$$
$$(g,f) \mapsto \left(H(g,f) - \fint_M H(g,f)\, dV_{g_E}, \int_M e^{-f}\, dV_g - 1\right).$$

Here, $C^{0,\alpha}_{g_E}(M) = \left\{f \in C^{0,\alpha}(M) \mid \int_M f\, dV_{g_E} = 0\right\}$. This is an analytic map between Banach manifolds. Observe that $L(g,f) = (0,0)$ if and only if we have $H(f,g) = const$ and $\int_M e^{-f}\, dV_g = 1$. The differential of L at (g_E, f_{g_E}) restricted to its second argument is equal to

$$dL_{g_E,f_{g_E}}(0,v) = \left(-(\Delta_{g_E}+1)v + \fint_M v\, dV, -\fint_M v\, dV\right).$$

The map $dL_{g_E,f_{g_E}}|_{C^{2,\alpha}(M)} : C^{2,\alpha}(M) \to C^{0,\alpha}_{g_E}(M) \times \mathbb{R}$ is a linear isomorphism because it acts as $-(\Delta_{g_E}+1)$ on $C^{2,\alpha}_{g_E}$ and as $-\mathrm{id}$ on constant functions. By the implicit function theorem for Banach manifolds, there exists a neighbourhood $\mathcal{U} \subset \mathcal{M}^{C^{2,\alpha}}$ and an analytic map $P\colon \mathcal{U} \to C^{2,\alpha}(M)$ such that we have $L(g, P(g)) = (0,0)$. Moreover, there exists a neighbourhood $\mathcal{V} \subset C^{2,\alpha}(M)$ of f_{g_E} such that if $L(g,f) = 0$ for some $g \in \mathcal{U}, f \in \mathcal{V}$, then $f = P(g)$.

Next, we show that $f_g = P(g)$ for all $g \in \mathcal{U}$ (or eventually on a smaller neighbourhood). Suppose this is not the case. Then there exists a sequence g_i which converges to g in $C^{2,\alpha}$ and such that $f_i \neq P(g_i)$ for all i. By the proof of Lemma 5.3.1, $\|f_{g_i}\|_{C^{2,\alpha}}$ is bounded and for every $\alpha' < \alpha$, there is a subsequence, again denoted by f_{g_i} converging to f_{g_E} in $C^{2,\alpha'}$. We obviously have $L(g_i, f_{g_i}) = (0,0)$ and for sufficiently large i we have, by the implicit function theorem, $f_{g_i} = P(g_i)$. This causes the contradiction.

We immediately get that $\mu_+(g) = H(g, P(g))$ is analytic on $\mathcal{U}$ since H and P are analytic. □

Theorem 5.5.3 (Lojasiewicz-Simon inequality for μ_+). *Let (M, g_E) be a Einstein manifold with constant -1. Then there exists a $C^{2,\alpha}$-neighbourhood $\mathcal{U}$ of g_E in the space of metrics and constants $\sigma \in [1/2, 1)$, $C > 0$ such that*

$$|\mu_+(g) - \mu_+(g_E)|^\sigma \le C \left\|\mathrm{Ric}_g + h + \nabla^2 f_g\right\|_{L^2} \tag{5.25}$$

for all $g \in \mathcal{U}$.

Proof. The proof is an application of a general Lojasiewicz-Simon inequality which was proven in [CM12]. Here the analyticity of μ_+ is crucial.

Since both sides are diffeomorphism invariant, it suffices to show the inequality on a slice to the action of the diffeomorphism group. Let

$$\mathcal{S}_{g_E} = \mathcal{U} \cap \left\{g_E + h \mid h \in \delta_{g_E}^{-1}(0)\right\},$$

and let $\tilde{\mu}_+$ be the μ_+-functional restricted to $\mathcal{S}_{g_E}$. Obviously, $\tilde{\mu}_+$ is analytic since μ_+ is. The L^2-gradient of μ_+ is given by $\nabla\mu_+(g) = -\frac{1}{2}(\mathrm{Ric}_g + g + \nabla^2 f_g)e^{-f_g}$. It vanishes at g_E. On the neighbourhood $\mathcal{U}$, we have the uniform estimate

$$\|\nabla\mu_+(g_1) - \nabla\mu_+(g_2)\|_{L^2} \le C_1 \|g_1 - g_2\|_{H^2}, \tag{5.26}$$

which holds by Taylor expansion and Lemma 5.3.2. The L^2-gradient of $\tilde{\mu}_+$ is given by the projection of $\nabla\mu_+$ to $\delta_{g_E}^{-1}(0)$. Therefore, (5.26) also holds for $\nabla\tilde{\mu}_+$. The linearization of $\tilde{\mu}_+$ at g_E is (up to a constant factor) given by the Einstein operator, see Lemma 5.2.5 (iii). By ellipticity,

$$\Delta_E\colon (\delta_{g_E}^{-1}(0))^{C^{2,\alpha}} \to (\delta_{g_E}^{-1}(0))^{C^{0,\alpha}}$$

is Fredholm. It also satisfies the estimate $\|\Delta_E h\|_{L^2} \le C_2 \|h\|_{H^2}$.

By [CM12, Theorem 6.3], there exists a constant $\sigma \in [1/2, 1)$ such that $|\mu_+(g) - \mu_+(g_E)|^\sigma \leq \|\nabla \tilde{\mu}_+(g)\|_{L^2}$ for any $g \in \mathcal{S}_{g_E}$. Since

$$\|\nabla \tilde{\mu}_+(g)\|_{L^2} \leq \|\nabla \mu_+(g)\|_{L^2} \leq C_3 \left\|\mathrm{Ric}_g + h + \nabla^2 f_g\right\|_{L^2},$$

(5.25) holds for all $g \in \mathcal{S}_{g_E}$. By diffeomorphism invariance, it holds for all $g \in \mathcal{U}$. □

Remark 5.5.4. Because the Lojasiewicz exponent $\sigma \in [1/2, 1)$, the convergence rate will be polynomially. Exponential convergence only holds if $\sigma = 1/2$.

5.5.3 Dynamical Stability and Instability

The proofs of the following stability/instability results are, up to some modifications, of the same nature as the proofs in the integrable case.

Theorem 5.5.5 (Dynamical stability modulo diffeomorphism). *Let (M, g_E) be an Einstein manifold with constant -1. Let $k \geq 3$. If g_E is a local maximizer of the Yamabe functional, then for every C^k-neighbourhood $\mathcal{U}$ of g_E, there exists a C^{k+2}-neighbourhood $\mathcal{V}$ such that the following holds:*

For any metric $g_0 \in \mathcal{V}$ there exists a 1-parameter family of diffeomorphisms φ_t such that for the Ricci flow $g(t)$ starting at g_0, the modified flow $\varphi_t^* g(t)$ stays in $\mathcal{U}$ for all time and converges to an Einstein metric g_∞ with constant -1 in $\mathcal{U}$ as $t \to \infty$. The convergence is of polynomial rate, i.e. there exist constants $C, \alpha > 0$ such that

$$\|\varphi_t^* g(t) - g_\infty\|_{C^k} \leq C(t+1)^{-\alpha}.$$

Proof. Without loss of generality, we may assume that $\mathcal{U} = \mathcal{B}^k_\epsilon$ and $\epsilon > 0$ is so small that Theorems 5.5.1 and 5.5.3 hold on $\mathcal{U}$.

By Lemma 5.4.10, we can choose a small neighbourhood $\mathcal{V}$ such that the Ricci flow starting at any metric $g \in \mathcal{V}$ stays in $\mathcal{B}^k_{\epsilon/4}$ up to time 1. Let $T \geq 1$ be the maximal time such that for any Ricci flow $g(t)$ starting in $\mathcal{V}$, there exists a family of diffeomorphisms φ_t such that the modified flow $\varphi_t^* g(t)$ stays in $\mathcal{U}$. By definition of T and diffeomorphism invariance, we have uniform curvature bounds

$$\sup_{p \in M} |R_{g(t)}|_{g(t)} \leq C_1 \qquad \forall t \in [0, T).$$

By Remark 5.4.12, we have

$$\sup_{p \in M} |\nabla^l R_{g(t)}|_{g(t)} \leq C(l) \qquad \forall t \in [1, T). \tag{5.27}$$

Because $f_{g(t)}$ satisfies the equation $-\Delta f_g - \frac{1}{2}|\nabla f_g|^2 + \frac{1}{2}\mathrm{scal}_g - f_g = \mu_+(g)$, we also have

$$\sup_{p \in M} |\nabla^l f_{g(t)}|_{g(t)} \leq \tilde{C}(l), \qquad \forall t \in [1, T). \tag{5.28}$$

Note that all these estimates are diffeomorphism invariant.

We now construct a modified Ricci flow as follows: Let $\varphi_t \in \mathrm{Diff}(M)$, $t \geq 1$ be the family of diffeomorphisms generated by $X(t) = -\mathrm{grad}_{g(t)} f_{g(t)}$ and define

$$\tilde{g}(t) = \begin{cases} g(t), & t \in [0,1], \\ \varphi_t^* g(t), & t \geq 1. \end{cases} \tag{5.29}$$

The modified flow satisfies the usual Ricci flow equation for $t \in [0,1]$ while for $t \geq 1$, we have

$$\begin{aligned} \frac{d}{dt}\tilde{g}(t) &= \varphi_t^*(\dot{g}(t)) + \varphi_t^*(\mathcal{L}_{X(t)} g(t)) \\ &= -2\varphi_t^*(\mathrm{Ric}_{g(t)} + g(t)) - 2\varphi_t^*(\nabla^2 f_{g(t)}) \\ &= -2(\mathrm{Ric}_{\tilde{g}(t)} + \tilde{g}(t) + \nabla^2 f_{\tilde{g}(t)}). \end{aligned}$$

Let $T' \in [0,T]$ be the maximal time such that the modified Ricci flow, starting at any metric $g_0 \in \mathcal{V}$, stays in $\mathcal{U}$ up to time T'. Then

$$\begin{aligned} \|\tilde{g}(T') - g_E\|_{C^k_{g_E}} &\leq \|\tilde{g}(1) - g_E\|_{C^k} + \int_1^{T'} \|\dot{\tilde{g}}(t)\|_{C^k_{g_E}} \, dt \\ &\leq \frac{\epsilon}{4} + 2\int_1^{T'} \|\dot{\tilde{g}}(t)\|_{C^k_{g(t)}} \, dt, \end{aligned}$$

provided that $\mathcal{U}$ is small enough. By the interpolation inequality for tensors (see [Ham82, Corollary 12.7]), (5.27) and (5.28), we have

$$\|\dot{\tilde{g}}(t)\|_{C^k_{g(t)}} \leq C_2 \|\dot{\tilde{g}}(t)\|_{L^2_{g(t)}}^{1-\eta}$$

for η as small as we want. In particular, we can assume that

$$\theta := 1 - \sigma(1+\eta) > 0,$$

where σ is the exponent appearing in Theorem 5.5.3. By the first variation of μ_+,

$$\frac{d}{dt}\mu_+(\tilde{g}(t)) \geq C_3 \|\dot{\tilde{g}}(t)\|_{L^2_{g(t)}}^{1+\eta} \|\dot{\tilde{g}}(t)\|_{L^2_{g(t)}}^{1-\eta}.$$

By Theorem 5.5.1 and Theorem 5.5.3 again,

$$\begin{aligned} -\frac{d}{dt}|\mu_+(\tilde{g}(t)) - \mu_+(g_E)|^\theta &= \theta|\mu_+(\tilde{g}(t)) - \mu_+(g_E)|^{\theta-1}\frac{d}{dt}\mu_+(\tilde{g}(t)) \\ &\geq C_4|\mu_+(\tilde{g}(t)) - \mu_+(g_E)|^{-\sigma(1+\eta)} \|\dot{\tilde{g}}(t)\|_{L^2_{g(t)}}^{1+\eta} \|\dot{\tilde{g}}(t)\|_{L^2_{g(t)}}^{1-\eta} \\ &\geq C_5 \|\dot{\tilde{g}}(t)\|_{C^k_{g(t)}}. \end{aligned}$$

Hence by integration,

$$\int_1^{T'} \|\dot{\tilde{g}}(t)\|_{C^k} \, dt \leq \frac{1}{C_5}|\mu_+(\tilde{g}(1)) - \mu_+(g_E)|^\theta \leq \frac{1}{C_5}|\mu_+(\tilde{g}(0)) - \mu_+(g_E)|^\theta \leq \frac{\epsilon}{8},$$

provided that $\mathcal{V}$ is small enough. This shows that $\|\tilde{g}(T') - g_E\|_{C^k_{g_E}} \leq \epsilon$, so T' cannot be finite. Thus, $T = \infty$ and $\tilde{g}(t)$ converges to some limit metric $g_\infty \in \mathcal{U}$ as $t \to \infty$. By the Lojasiewicz-Simon inequality, we have

$$\frac{d}{dt}|\mu_+(\tilde{g}(t)) - \mu_+(g_E)|^{1-2\sigma} \geq C_6,$$

which implies

$$|\mu_+(\tilde{g}(t)) - \mu_+(g_E)| \leq C_7(t+1)^{-\frac{1}{2\sigma-1}}.$$

Here, we may assume that $\sigma > \frac{1}{2}$ because the Lojasiewicz-Simon inequality also holds after enlarging the exponent. Therefore, $\mu_+(g_\infty) = \mu_+(g_E)$, so g_∞ is an Einstein metric with constant -1. The convergence is of polynomial rate since for $t_1 < t_2$,

$$\|\tilde{g}(t_1) - \tilde{g}(t_2)\|_{C^k} \leq C_8|\mu_+(\tilde{g}(t_1)) - \mu_+(g_E)|^\theta \leq C_9(t_1+1)^{-\frac{\theta}{2\sigma-1}},$$

and the assertion follows from $t_2 \to \infty$. □

Theorem 5.5.6 (Dynamical instability modulo diffeomorphism). *Let (M, g_E) be an Einstein manifold with constant -1 which is not a local maximizer of the Yamabe functional. Then there exists an ancient Ricci flow $g(t)$, defined on $(-\infty, 0]$, and a 1-parameter family of diffeomorphisms φ_t, $t \in (-\infty, 0]$ such that $\varphi_t^* g(t) \to g_E$ as $t \to -\infty$.*

Proof. Since (M, g_E) is not a local maximum of the Yamabe functional, it cannot be a local maximum of μ_+. Let $g_i \to g_E$ in C^k and suppose that we have $\mu_+(g_i) > \mu_+(g_E)$ for all i. Let $\tilde{g}_i(t)$ be the modified flow defined in (5.29), starting at g_i. Then by Lemma 5.4.10, $\bar{g}_i = g_i(1)$ converges to g_E in C^{k-2} and by monotonicity, $\mu_+(\bar{g}_i) > \mu_+(g_E)$ as well. Let $\epsilon > 0$ be so small that Theorem 5.5.3 holds on $\mathcal{B}^{k-2}_{2\epsilon}$. Theorem 5.5.3 yields the differential inequality

$$\frac{d}{dt}(\mu_+(\tilde{g}_i(t)) - \mu_+(g_E))^{1-2\sigma} \geq -C_1,$$

from which we obtain

$$[(\mu_+(\tilde{g}_i(t)) - \mu_+(g_E))^{1-2\sigma} - C_1(s-t)]^{-\frac{1}{2\sigma-1}} \leq (\mu_+(\tilde{g}_i(s)) - \mu_+(g_E))$$

as long as $\tilde{g}_i(t)$ stays in $\mathcal{B}^{k-2}_{2\epsilon}$. Thus, there exists a t_i such that

$$\|\tilde{g}_i(t_i) - g_E\|_{C^{k-2}} = \epsilon,$$

and $t_i \to \infty$. If $\{t_i\}$ was bounded, $\tilde{g}_i(t_i) \to g_E$ in C^{k-2}. By interpolation,

$$\left\|\mathrm{Ric}_{\tilde{g}_i(t)} - \tilde{g}_i(t)\right\|_{C^{k-2}} \leq C_2 \left\|\mathrm{Ric}_{\tilde{g}_i(t)} - \tilde{g}_i(t)\right\|_{L^2}^{1-\eta}$$

for $\eta > 0$ as small as we want. We may assume that $\theta = 1 - \sigma(1+\eta) > 0$. By Theorem 5.5.3 , we have the differential inequality

$$\frac{d}{dt}(\mu_+(\tilde{g}_i(t)) - \mu_+(g_E))^\theta \geq C_3 \left\|\mathrm{Ric}_{\tilde{g}_i(t)} + \tilde{g}_i(t)\right\|_{L^2}^{1-\eta},$$

if $\mu_+(\tilde{g}_i(t)) > \mu_+(g_E)$. Thus,

$$\epsilon = \|\tilde{g}_i(t_i) - g_E\|_{C^{k-2}} \leq \|\bar{g}_i - g_E\|_{C^{k-2}} + C_4(\mu_+(\tilde{g}_i(t_i)) - \mu_+(g_E))^\theta. \quad (5.30)$$

Now put $\tilde{g}_i^s(t) := \tilde{g}_i(t+t_i)$, $t \in [T_i, 0]$, where $T_i = 1 - t_i \to -\infty$. We have

$$\|\tilde{g}_i^s(t) - g_E\|_{C^{k-2}} \leq \epsilon \qquad \forall t \in [T_i, 0],$$
$$\tilde{g}_i^s(T_i) \to g_E \text{ in } C^{k-2}.$$

Because the embedding $C^{k-3}(M) \subset C^{k-2}(M)$ is compact, we can choose a subsequence of the $\tilde{g}_i^s$, converging in $C^{k-3}_{loc}(M \times (-\infty, 0])$ to an ancient flow $\tilde{g}(t)$, $t \in (-\infty, 0]$, satisfying the differential equation

$$\dot{\tilde{g}}(t) = -2(\mathrm{Ric}_{\tilde{g}(t)} + \tilde{g}(t) + \nabla^2 f_{\tilde{g}(t)}).$$

Let φ_t, $t \in (-\infty, 0]$ be the diffeomorphisms generated by $X(t) = \mathrm{grad}_{\tilde{g}(t)} f_{\tilde{g}}$ where $\varphi_0 = \mathrm{id}$. Then $g(t) = \varphi_t^* \tilde{g}(t)$ is a solution of (5.5). From taking the limit $i \to \infty$ in (5.30), we obtain $\epsilon \leq C_4(\mu_+(g(0)) - \mu_+(g_E))^{\beta/2}$ and therefore, the Ricci flow is nontrivial. For $T_i \leq t$, we have, by the Lojasiewicz-Simon inequality,

$$\begin{aligned}\|\tilde{g}_i^s(T_i) - \tilde{g}_i^s(t)\|_{C^{k-3}} \leq & C_4(\mu_+(\tilde{g}_i(t+t_i)) - \mu_+(g_E))^\theta \\ \leq & C_4[-C_1 t + (\mu_+(\tilde{g}_i(t_i)) - \mu_+(g_E))^{1-2\sigma}]^{-\frac{\theta}{2\sigma-1}} \\ \leq & [-C_5 t + C_6]^{-\frac{\theta}{2\sigma-1}}.\end{aligned}$$

Thus,

$$\begin{aligned}\|g_E - \tilde{g}(t)\|_{C^{k-3}} \leq & \|g_E - \tilde{g}_i^s(T_i)\|_{C^{k-3}} + [-C_5 t + C_6]^{-\frac{\theta}{2\sigma-1}} \\ & + \|\tilde{g}_i^s(t) - \tilde{g}(t)\|_{C^{k-3}}.\end{aligned}$$

It follows that $\|g_E - \tilde{g}(t)\|_{C^{k-3}} \to 0$ as $t \to -\infty$. Therefore, $(\varphi_t^{-1})^* g(t) \to g_E$ in C^{k-3} as $t \to -\infty$. □

Remark 5.5.7. The previous theorems in particular imply the following: Any compact negative Einstein metric is either dynamically stable or dynamically unstable modulo diffeomorphism and this only depends on the local behaviour of the Yamabe functional.

On manifolds with Yamabe invariant $Y(M) \leq 0$, it is well-known that any metric realizing the Yamabe invariant is Einstein (see e.g. [And05]). From Theorem 5.1.4, Remark 5.1.5 and Theorem 5.5.5, we thus obtain

Corollary 5.5.8. *Let M be a manifold with $Y(M) \leq 0$. Then any metric on M realizing the Yamabe invariant is a dynamically stable Einstein manifold.*

Chapter 6

Ricci Flow and positive Einstein Metrics

6.1 Introduction

In this chapter we prove analogous stability and instability results to those in Chapter 5 for positive Einstein manifolds.

To deal with positive Einstein manifolds we use a different variant of the Ricci flow which is defined by the differential equation

$$\dot{g}(t) = -2\mathrm{Ric}_{g(t)} + \frac{2}{n}\left(⨍_M \mathrm{scal}_{g(t)}\, dV_{g(t)}\right) g(t). \tag{6.1}$$

It has the property that the volume is preserved under the flow. If $g(t)$ be a solution of (6.1), then

$$\begin{aligned} \frac{d}{dt}\mathrm{vol}(M, g(t)) =& \frac{1}{2}\int_M \mathrm{tr}\dot{g}(t)\, dV_{g(t)} \\ =& \int_M \left[-\mathrm{scal}_{g(t)} + \left(⨍_M \mathrm{scal}_{g(t)}\, dV_{g(t)}\right)\right] dV_{g(t)} = 0. \end{aligned}$$

We now translate the definition of dynamical stability and instability to positive Einstein metrics and with respect to (6.1).

Definition 6.1.1 (Dynamical stability and instability). Let (M, g_E) be a positive Einstein manifold. We call (M, g_E) dynamically stable if for every neighbourhood $\mathcal{U}$ of g_E in the space of metrics there exists a smaller neighbourhood $\mathcal{V} \subset \mathcal{U}$ such that the Ricci flow (6.1) starting in $\mathcal{V}$ stays in $\mathcal{U}$ for all $t \geq 0$ and converges to an Einstein metric as $t \to \infty$.

We call (M, g_E) dynamically stable modulo diffeomorphism if for each solution of (6.1) starting in $\mathcal{V}$, there exists a familiy of diffeomorphisms φ_t, $t \geq 0$ such that the modified flow $\varphi_t^* g(t)$ stays in $\mathcal{U}$ for all $t \geq 0$ and converges to an Einstein metric as $t \to \infty$.

We call (M, g_E) dynamically unstable (modulo diffeomorphism) if there exists an ancient flow $g(t)$, $t \in (-\infty, T]$ such that $g(t) \to g_E$ as $t \to -\infty$ (there exists a family of diffeomorphisms φ_t, $t \in (-\infty, T]$ such that $\varphi_t^* g(t) \to g_E$ as $t \to -\infty$).

Remark 6.1.2. The reason why we do not deal with the flow

$$\dot{g}(t) = -2(\mathrm{Ric}_{g(t)} - g(t)), \tag{6.2}$$

which is the natural analogue of (5.5), is that its stationary points are never dynamically stable. Let $c \in \mathbb{R}$. If g_E is an Einstein manifold with constant 1, the solution of (6.2) starting at $(1+c)g_E$ is given by $(1+ce^{2t})g_E$, which clearly diverges as long as $c \neq 0$.

It is well known that the standard sphere is dynamically stable. This was proven in [Ham82] for $n = 3$ and in [Hui85] for $n \geq 4$. We have also already seen that it is Einstein-Hilbert stable. Therefore we assume throughout the chapter that $(M, g) \neq (S^n, g_{st})$.

The stability/instability conditions are quite simliar as in the negative case. However, the situation is slightly more subtle because a condition on the spectrum of the Laplace operator comes into play as we will see.

6.2 The Shrinker Entropy

We define the Ricci shrinker entropy which was first introduced by G. Perelman in [Per02]. Let

$$\mathcal{W}_-(g, f, \tau) = \frac{1}{(4\pi\tau)^{n/2}} \int_M [\tau(|\nabla f|_g^2 + \mathrm{scal}_g) + f - n]e^{-f}\, dV.$$

For $\tau > 0$, let

$$\mu_-(g, \tau) = \inf\left\{\mathcal{W}_-(g, f, \tau) \,\middle|\, f \in C^\infty(M), \frac{1}{(4\pi\tau)^{n/2}} \int_M e^{-f}\, dV_g = 1\right\}.$$

For any $\tau > 0$, the infimum is realized by a smooth function. We define the shrinker entropy as

$$\nu_-(g) = \inf\{\mu_-(g, \tau) \mid \tau > 0\}.$$

If $\lambda(g) > 0$ (see (5.2) for the definition), then $\nu_-(g)$ is finite and realized by some $\tau_g > 0$ (see [CCG$^+$07, Corollary 6.34]). In this case, a pair (f_g, τ_g) realizing $\nu_-(g)$ satisfies the equations

$$\tau(2\Delta f + |\nabla f|^2 - \mathrm{scal}) - f + n + \nu_- = 0, \tag{6.3}$$

$$\frac{1}{(4\pi\tau)^{n/2}} \int_M f e^{-f}\, dV = \frac{n}{2} + \nu_-, \tag{6.4}$$

see e.g. [CZ12, p. 5].

Remark 6.2.1. Note that $\mathcal{W}_-(\varphi^* g, \varphi^* f, \tau) = \mathcal{W}_-(g, f, \tau)$ for $\varphi \in \mathrm{Diff}(M)$ and $\mathcal{W}_-(\alpha g, f, \alpha\tau) = \mathcal{W}_-(g, f, \tau)$ for $\alpha > 0$. Therefore, $\nu_-(g) = \nu_-(\alpha \cdot \varphi^* g)$ for any $\varphi \in \mathrm{Diff}(M)$ and $\alpha > 0$.

Proposition 6.2.2 (First variation of ν_-). *Let (M, g) be a Riemannian manifold. Then the first variation of ν_- is given by*

$$\nu_-(g)'(h) = -\frac{1}{(4\pi\tau)^{n/2}} \int_M \left\langle \tau_g(\mathrm{Ric} + \nabla^2 f_g) - \frac{1}{2}g, h \right\rangle e^{-f_g}\, dV_g,$$

where (f_g, τ_g) realizes $\nu_-(g)$. Consequently, ν_- is nondecreasing under any solution of (6.1).

Proof. A proof of the first variational formula is given in many papers, see e.g. [CZ12, Lemma 2.2]. By scale and diffeomorphism invariance,

$$\nu_-(g)'(\nabla^2 f_g) = \nu_-(g)'(\mathcal{L}_{\mathrm{grad}f}g) = 0,$$
$$\nu_-(g)'\left(\left(\frac{1}{2\tau} - \frac{1}{n}\fint_M \mathrm{scal}\cdot\, dV\right)\cdot g\right) = 0.$$

Therefore, if $g(t)$ is a solution of (6.1), the time-derivative of $\nu_-(g(t))$ is equal to

$$\frac{2\tau_{g(t)}}{(4\pi\tau_{g(t)})^{n/2}}\int_M \left|\mathrm{Ric}_{g(t)} + \nabla^2 f_{g(t)} - \frac{1}{2\tau_{g(t)}}g(t)\right|^2 e^{-f_{g(t)}}\, dV_{g(t)} \geq 0, \tag{6.5}$$

which finishes the proof. □

The critical metrics of ν_- are those which satisfy $\mathrm{Ric} + \nabla^2 f_g - \frac{1}{2\tau}g = 0$. We call such metrics shrinking gradient Ricci solitons. Any positive Einstein metric (M, g_E) is a stationary point of (6.1). Since equality must hold in (6.5), it is a shrinking gradient Ricci soliton and because f_{g_E} is nessecarily constant, the pair (f_{g_E}, τ_{g_E}) satisfies

$$\tau_{g_E} = \frac{1}{2\mu}, \qquad f_{g_E} = \log(\mathrm{vol}(M, g_E)) - \frac{n}{2}(\log(2\pi) - \log(\mu)), \tag{6.6}$$

where μ is the Einstein constant.

Lemma 6.2.3. *Let (M, g_E) be a positive Einstein metric with constant μ. Then*

(i) $\frac{d}{dt}|_{t=0}\tau_{g_E+th} = \frac{\tau}{n}\fint \mathrm{tr}h\; dV$.

If $\delta h = 0$ and $\int_M \mathrm{tr}h\; dV = 0$, then

(ii) $\frac{d}{dt}|_{t=0} f_{g_E+th} = \frac{1}{2}\mathrm{tr}h$,

(iii) $\frac{d}{dt}|_{t=0}(\tau_{g_E+th}(\mathrm{Ric}_{g_E+th} + \nabla^2 f_{g_E+th}) - \frac{1}{2}(g_E + th)) = \frac{1}{4\mu}\Delta_E h$,

where Δ_E is the Einstein operator.

Proof. The first variation of τ at shrinking gradient Ricci solitons was computed by Cao and Zhu (see [CZ12, Lemma 2.4]). It is given by

$$\left.\frac{d}{dt}\right|_{t=0}\tau_{g+th} = \tau_g\frac{\int_M \langle\mathrm{Ric}, h\rangle e^{-f_g}\, dV}{\int_M \mathrm{scal}\cdot e^{-f_g}\, dV}.$$

This is (i) in the case of positive Einstein metrics. To compute (ii), we differentiate equation (6.3) at g_E and we obtain

$$\tau(2\Delta f' - \mathrm{scal}') - \tau'\mathrm{scal} - f' = 0.$$

Since $\int_M \mathrm{tr}h\; dV = 0$, τ' vanishes and since $\delta h = 0$,

$$\frac{1}{\mu}\Delta f' - f' = \tau\mathrm{scal}' = \tau(\Delta(\mathrm{tr}h) - \mu\mathrm{tr}h) = \frac{1}{2\mu}(\Delta\mathrm{tr}h - \mu\mathrm{tr}h).$$

By Obata's eigenvalue estimate, $\mu\Delta - 1$ is invertible and (ii) follows. The proof of (iii) is done by straightforward computation. By using $\delta h = 0$, $\tau' = 0$ and (ii),

$$\begin{aligned}(\tau(\mathrm{Ric} + \nabla^2 f) - \frac{1}{2}g)' &= \tau(\mathrm{Ric}' + \nabla^2 f') - \frac{1}{2}h \\ &= \tau\left(\frac{1}{2}\Delta_L h - \delta^*(\delta h) - \frac{1}{2}\nabla^2 \mathrm{tr} h + \frac{1}{2}\nabla^2 \mathrm{tr} h\right) - \frac{1}{2}h \\ &= \frac{1}{2\mu}\frac{1}{2}\Delta_L h - \frac{1}{2}h \\ &= \frac{1}{4\mu}(\Delta_L h - 2\mu h) = \frac{1}{4\mu}\Delta_E h. \qquad \square\end{aligned}$$

Before we continue computing the second variation on Einstein manifolds, we first remark that the splitting of $\delta_{g_E}^{-1}(0) \subset \Gamma(S^2 M)$ proven in Lemma 5.4.2 can be refined to

$$\delta_{g_E}^{-1}(0) = \mathbb{R} \cdot g_E \oplus C_{g_E}(C^\infty_{g_E}(M)) \oplus TT_{g_E},$$

and this splitting is again orthogonal. Recall that $C_{g_E}(f) = (\Delta f - \mu f)g_E + \nabla^2 f$ and that $C^\infty_{g_E}(M)$ denotes the space of smooth functions with $\int_M f \, dV_{g_E} = 0$. The whole space of symmetric $(0,2)$-tensor fields splits orthogonally as

$$\Gamma(S^2 M) = \delta^*(\Omega^1(M)) \oplus \mathbb{R} \cdot g_E \oplus C_{g_E}(C^\infty_{g_E}(M)) \oplus TT_{g_E}.$$

The second variation of ν_- on shrinking gradient Ricci solitons was already computed in [CHI04; CZ12]. The decomposition above allows us to state it in a simpler form. Moreover, the formula simplifies because we only treat the Einstein case.

Proposition 6.2.4 (Second variation of ν_-). *The second variation of ν_- on a postive Einstein metric (M, g_E) with constant μ is given by*

$$\nu_-(g_E)''(h) = \begin{cases} -\frac{1}{4\mu} ⨍_M \langle \Delta_E h, h\rangle \, dV, & \text{if } h \in C_{g_E}(C^\infty_{g_E}(M)) \oplus TT_{g_E}, \\ 0, & \text{if } h \in \mathbb{R} \cdot g_E \oplus \delta^*(\Omega^1(M)). \end{cases}$$

Proof. By scale and diffeomorphism invariance, $\nu_-(g_E)''$ vanishes on the subspace $\mathbb{R} \cdot g_E \oplus \delta^*(\Omega^1(M))$. Now let $h \in C_{g_E}(C^\infty_{g_E}(M)) \oplus TT_{g_E}$. Note that $\delta h = 0$ and $\int_M \mathrm{tr} h \, dV = 0$. By Lemma 6.2.3 (iii),

$$\begin{aligned}\frac{d^2}{dt^2}\bigg|_{t=0} \nu_-(g_E + th) &= -\frac{d}{dt}\bigg|_{t=0} \frac{1}{(4\pi\tau)^{n/2}} \int_M \left\langle \tau(\mathrm{Ric} + \nabla^2 f_g) - \frac{1}{2}g, h\right\rangle e^{-f_g} \, dV \\ &= -\frac{1}{(4\pi\tau)^{n/2}} \int_M \left\langle \frac{d}{dt}\bigg|_{t=0} (\tau(\mathrm{Ric} + \nabla^2 f_g) - \frac{1}{2}g), h\right\rangle e^{-f_g} \, dV \\ &= -\frac{1}{4\mu} ⨍_M \langle \Delta_E h, h\rangle \, dV.\end{aligned}$$

Moreover, since the Einstein operator and the Lichnerowicz Laplacian satisfy the relation $\Delta_L = \Delta_E + 2\mu \cdot \mathrm{id}$, Lemma 2.4.5 implies that the Einstein operator preserves the subspaces $C_{g_E}(C^\infty_{g_E}(M))$ and TT_{g_E}. Thus, the splitting of above is orthogonal with respect to $\nu_-(g_E)''$. $\square$

Corollary 6.2.5. *Let (M, g_E) be a positive Einstein manifold with constant μ. Then dynamical stability (modulo diffeomorphism) implies Einstein-Hilbert stability. Moreover, it implies that the smallest nonzero eigenvalue of the Laplacian satisfies the bound $\lambda \geq 2\mu$.*

Proof. We have seen that ν_- is nondecreasing under (6.1) and that ν_- is invariant under diffeomorphisms. Thus, we have that (M, g_E) is nessecarily a local maximum of ν_-, if it is dynamically stable (modulo diffeomorphism). The second variational formula implies that the Einstein operator is nonnegative on $C_{g_E}(C^\infty_{g_E}(M)) \oplus TT_{g_E}$. Einstein-Hilbert stability follows from definition. Moreover, for any $f \in C^\infty_{g_E}(M)$, Lemma 2.4.5 implies

$$\Delta_E(C_{g_E}(f)) = (\Delta_L - 2\mu)(C_{g_E}(f)) = C_{g_E}((\Delta - 2\mu)f).$$

Since we excluded the case $(M, g) = (S^n, g_{st})$, we conclude from Lemma 2.4.1 that $C_{g_E} \colon C^\infty_{g_E}(M) \to \Gamma(S^2M)$ is injective. Thus, $\Delta - 2\mu$ is nonnegative on $C^\infty_{g_E}(M)$, which proves the eigenvalue bound. □

Definition 6.2.6. An Einstein manifold (M, g_E) is called linearly stable if $\nu_-(g_E)''$ is negative semidefinite. A linearly stable Einstein manifold is called neutrally linearly stable if $\nu_-(g_E)''(h) = 0$ for some $h \in C_{g_E}(C^\infty_{g_E}(M)) \oplus TT_{g_E}$.

6.3 Some technical Estimates

This section contains similar technical estimates to those in Section 5.3. We prove estimates on $\nu_-(g), f_g, \tau_g$ and their variations in terms of norms of the variations. Compared to Section 5.3, more technical effort is needed because we have to deal with a coupled pair of Euler-Lagrange equations satisfied by (f_g, τ_g).

Lemma 6.3.1. *Let (M, g_E) be a compact Einstein manifold. Then there exists a $C^{2,\alpha}$-neighbourhood $\mathcal{U}$ of g_E such that the minimizing pair (f_g, τ_g) realizing $\nu_-(g)$ is unique and depends analytically on the metric. Moreover, the map $g \mapsto \nu_-(g)$ is analytic on $\mathcal{U}$.*

Proof. We again use an implicit function argument. We define a map H by $H(g, f, \tau) = \tau(2\Delta f + |\nabla f|^2 - \mathrm{scal}) - f + n$. Let

$$C^{k,\alpha}_{g_E}(M) = \left\{ f \in C^{k,\alpha}(M) \,\middle|\, \int_M f \, dV_{g_E} = 0 \right\}.$$

Define

$$\begin{aligned} L \colon \mathcal{M}^{C^{2,\alpha}} \times C^{2,\alpha}(M) \times \mathbb{R}_+ &\to C^{0,\alpha}_{g_E}(M) \times \mathbb{R} \times \mathbb{R}, \\ (g, f, \tau) &\mapsto (L_1, L_2, L_3), \end{aligned}$$

where the three components are given by

$$\begin{aligned} L_1(g, f, \tau) &= H(g, f, \tau) - \fint_M H(g, f, \tau) \, dV_{g_E}, \\ L_2(g, f, \tau) &= \frac{1}{(4\pi\tau)^{n/2}} \int_M f e^{-f} \, dV_g - \frac{n}{2} + \fint_M H(g, f, \tau) \, dV_g, \\ L_3(g, f, \tau) &= \frac{1}{(4\pi\tau)^{n/2}} \int_M e^{-f} \, dV_g - 1. \end{aligned}$$

This is an analytic map between Banach manifolds. We have $L(g,f,\tau)=(0,0,0)$ if and only if there exists a $c\in\mathbb{R}$ such that the set of equations

$$\tau(2\Delta f+|\nabla f|^2-\mathrm{scal})-f+n=c, \tag{6.7}$$

$$\frac{1}{(4\pi\tau)^{n/2}}\int_M fe^{-f}\,dV-\frac{n}{2}=-c, \tag{6.8}$$

$$\frac{1}{(4\pi\tau)^{n/2}}\int_M e^{-f}\,dV=1 \tag{6.9}$$

is satisfied. Now we compute the differential of L at $(g_E, f_{g_E}, \tau_{g_E})$ restricted to $R=C^{2,\alpha}(M)\times\mathbb{R}$. We use the splitting $C^{2,\alpha}(M)=C^{2,\alpha}_{g_E}(M)\times\mathbb{R}$ via $f\mapsto(f-\fint_M f\,dV_{g_E},\fint_M f\,dV_{g_E})$ to write

$$dL_{(g_E,f_{g_E},\tau_{g_E})}|_R\colon C^{2,\alpha}_{g_E}(M)\times\mathbb{R}\times\mathbb{R}\to C^{0,\alpha}_{g_E}(M)\times\mathbb{R}\times\mathbb{R}$$

as the matrix

$$dL_{(g_E,f_{g_E},\tau_{g_E})}|_R=\begin{pmatrix}\frac{1}{\mu}\Delta-1 & 0 & 0\\ 0 & -f_{g_E} & -n\mu(f_{g_E}+1)\\ 0 & -1 & -n\mu\end{pmatrix}.$$

Here, we used (6.6) to compute the matrix. This is an isomorphism since the map $\frac{1}{\mu}\Delta-1: C^{2,\alpha}_{g_E}(M)\to C^{0,\alpha}_{g_E}(M)$ is an isomorphism and the determinant of the 2×2-block is equal to $-n\mu\neq 0$. By the implicit function theorem for Banach manifolds, there exists a neighbourhood $\mathcal{U}\subset\mathcal{M}^{C^{2,\alpha}}$ of g_E and an analytic map $P\colon\mathcal{U}\to C^{2,\alpha}(M)\times\mathbb{R}_+$ such that $L(g,P(g))=0$. Moreover, there exists a neighbourhood $\mathcal{V}\subset C^{2,\alpha}(M)\times\mathbb{R}_+$ such that for any $(g,f,\tau)\in\mathcal{U}\times\mathcal{V}$, we have $L(g,f,\tau)=0$ if and only if $(f,\tau)=P(g)$.

Now, we prove that on a smaller neighbourhood $\mathcal{U}_1\subset\mathcal{U}$, there is a unique pair of minimizers in the definition of ν_- and it is equal to $P(g)$. Suppose this is not the case. Then there exist a sequence g_i of metrics such that $g_i\to g_E$ in $C^{2,\alpha}$ and pairs of minimizers (f_{g_i},τ_{g_i}) such that $P(g_i)\neq(f_i,\tau_{g_i})$ for all $i\in\mathbb{N}$. By substituting $w_{g_i}^2=e^{-f_{g_i}}$, we see that the pair (w_{g_i},τ_{g_i}) is a minimizer of the functional

$$\tilde{\mathcal{W}}_-(g_i,w,\tau)=\frac{1}{(4\pi\tau)^{n/2}}\int_M[\tau(4|\nabla w|^2+\mathrm{scal}_g w^2)-\log(w^2)w^2-nw^2]\,dV_{g_i}$$

under the constraint $\frac{1}{(4\pi\tau)^{n/2}}\int_M w^2\,dV_{g_i}=1$. It satisfies the pair of equations

$$-\tau_{g_i}(4\Delta w_{g_i}+\mathrm{scal}_{g_i}w_{g_i})-2\log(w_{g_i})w_{g_i}+nw_{g_i}+\nu_-(g_i)w_{g_i}=0, \tag{6.10}$$

$$-\frac{1}{(4\pi\tau_{g_i})^{n/2}}\int_M w_{g_i}^2\log w_{g_i}^2\,dV_{g_i}=\frac{n}{2}+\nu_-(g_i). \tag{6.11}$$

We have an upper bound $\nu_-(g_i)\leq C_1$ by testing with suitable pairs (f,τ). In fact, by choosing $f=\log(\mathrm{vol}(M,g_i)\cdot(\frac{2\pi}{\mu})^{-n/2})$ and $\tau=\frac{1}{2\mu}$, where μ is the Einstein constant of g_E, we have

$$\nu_-(g_i)\leq\frac{1}{2\mu}\sup\mathrm{scal}_{g_i}-n+\log\left(\mathrm{vol}(M,g_i)\cdot\left(\frac{2\pi}{\mu}\right)^{-n/2}\right). \tag{6.12}$$

Therefore,

$$\frac{1}{(4\pi\tau_{g_i})^{n/2}}\int_M [\tau_{g_i}(4|\nabla w_{g_i}|^2 + \mathrm{scal}_{g_i} w_{g_i}^2) - \log(w_{g_i}^2)w_{g_i}^2 - nw_{g_i}^2]\, dV_{g_i} \leq C_1.$$

Now, we show that there exist constants $C_2, C_3 > 0$ such that $C_2 \leq \tau_{g_i} \leq C_3$. Suppose this is not the case. By [CCG+07, Lemma 6.30], we have a lower estimate

$$\begin{aligned}\nu_-(g_i) = \mu_-(g_i, \tau_{g_i}) &\geq (\tau_{g_i} - 1)\lambda(g_i) - \frac{n}{2}\log\tau_{g_i} - C_4(g_i)\\ &\geq (\tau_{g_i} - 1)\inf \mathrm{scal}_{g_i} - \frac{n}{2}\log\tau_{g_i} - C_6\\ &\geq (\tau_{g_i} - 1)C_5 - \frac{n}{2}\log\tau_{g_i} - C_6.\end{aligned}$$

Here λ is the functional defined in (5.2), $C_5 > 0$ and $C_4(g)$ depends on the Sobolev constant and the volume. Now if τ_{g_i} converges to 0 or ∞, $\nu_-(g_i)$ diverges, which causes the contradiction. Observe that we also obtained a lower bound on $\nu_-(g_i)$.

Next, we show that $\|\nabla w_{g_i}\|_{L^2}$ is bounded. Choose $\epsilon > 0$ so small that $2 + 2\epsilon \leq \frac{2n}{n-2}$. By Jensen's inequality and the bounds on τ_{g_i},

$$\begin{aligned}\int_M w_{g_i}^2 \log w_{g_i}^2\, dV_{g_i} =& \frac{1}{\epsilon}\int_M w_{g_i}^2 \log w_{g_i}^{2\epsilon}\, dV_{g_i}\\ \leq& \frac{1}{\epsilon}\|w_{g_i}\|_{L^2}^2 \log\left(\frac{1}{\|w_{g_i}\|_{L^2}^2}\int_M w_{g_i}^{2+2\epsilon}\, dV_{g_i}\right)\\ =& \frac{1}{\epsilon}(4\pi\tau_{g_i})^{n/2}\log\left((4\pi\tau_{g_i})^{-n/2}\int_M w_{g_i}^{2+2\epsilon}\, dV_{g_i}\right)\\ \leq& C_7\log\left(\int_M w_{g_i}^{2+2\epsilon}\, dV_{g_i}\right) + C_8.\end{aligned}$$

By the Sobolev inequality,

$$\begin{aligned}\int_M w_{g_i}^{2+2\epsilon}\, dV_{g_i} \leq& C_9(\|\nabla w_{g_i}\|_{L^2}^2 + \|w_{g_i}\|_{L^2}^2)^{1+\epsilon}\\ \leq& C_9(\|\nabla w_{g_i}\|_{L^2}^2 + C_{10})^{1+\epsilon}.\end{aligned}$$

In summary, we have

$$\begin{aligned}C_1 &\geq \frac{1}{(4\pi\tau)^{n/2}}\int_M [\tau(4|\nabla w_{g_i}|^2 + \mathrm{scal}_{g_i} w_{g_i}^2) - \log(w_{g_i}^2)w_{g_i}^2 - nw^2]\, dV_{g_i}\\ &\geq C_{11}\|\nabla w_{g_i}\|_{L^2}^2 - C_{12}\log(\|\nabla w_{g_i}\|_{L^2}^2 + C_{10}) - C_{13},\end{aligned}$$

which shows that $\|\nabla w_{g_i}\|_{L^2}$ is bounded.

Now we proceed with a bootstrap argument similar to the proof of Lemma 5.3.1. By Sobolev embedding, the bound on $\|w_{g_i}\|_{H^1}$ implies a bound on $\|w_{g_i}\|_{L^{2n/(n-2)}}$. Let $p = 2n/(n-2)$ and choose some q slightly smaller than p. By elliptic regularity and (6.10),

$$\|w_{g_i}\|_{W^{2,q}} \leq C_{14}(\|w_{g_i}\log w_{g_i}\|_{L^q} + \|w_{g_i}\|_{L^q}).$$

Since for any $\beta > 1$, $|x \log x| \leq |x|^\beta$ for $|x|$ large enough, we have

$$\|w_{g_i} \log w_{g_i}\|_{L^q} \leq C_{15}(\mathrm{vol}(M, g_i)) + \|w_{g_i}\|_{L^p} \leq C_{16} + \|w_{g_i}\|_{L^p}.$$

Thus, $\|w_{g_i}\|_{W^{2,q}} \leq C(q)$. Using Sobolev embedding, we obtain bounds on $\|w_{g_i}\|_{L^{p'}}$ for some $p' > p$. From the (6.10) again, we have bounds on $\|w_{g_i}\|_{W^{2,q'}}$ for any $q' < p'$. Using this arguments repetitively, we obtain $\|w_{g_i}\|_{W^{2,q}} \leq C(q)$ for all $q \in (1, \infty)$. Again by elliptic regularity,

$$\begin{aligned}\|w_{g_i}\|_{C^{2,\alpha}} &\leq C_{17}(\|w_{g_i} \log w_{g_i}\|_{C^{0,\alpha}} + \|w_{g_i}\|_{C^{0,\alpha}})\\ &\leq C_{18}((\|w_{g_i}\|_{C^{0,\alpha}})^\gamma + \|w_{g_i}\|_{C^{0,\alpha}})\end{aligned}$$

for some $\gamma > 1$. For some sufficiently large q, we have, by Sobolev embedding,

$$\|w_{g_i}\|_{C^{0,\alpha}} \leq C_{19} \|w_{g_i}\|_{W^{1,q}} \leq C_{19} \cdot C(q).$$

We finally obtained an upper bound on $\|w_{g_i}\|_{C^{2,\alpha}}$. Thus, there exists a subsequence, again denoted by (w_{g_i}, τ_{g_i}), which converges in $C^{2,\alpha'}$, $\alpha' < \alpha$, to some limit (w_∞, τ_∞). We have, by (6.12),

$$\nu_-(g_E) \geq \lim_{i \to \infty} \nu_-(g_i) = \lim_{i \to \infty} \tilde{\mathcal{W}}_-(g_i, w_{g_i}, \tau_{g_i}) = \tilde{\mathcal{W}}_-(g_E, w_\infty, \tau_\infty) \geq \nu_-(g_E),$$

and therefore, $(w_\infty, \tau_\infty) = (w_{g_E}, \tau_{g_E})$ because the minimizing pair is unique at g_E by (6.6). Moreover, by resubstituting,

$$(f_{g_i}, \tau_{g_i}) \to (f_\infty, \tau_\infty) = (f_{g_E}, \tau_{g_E})$$

in $C^{2,\alpha'}$. Because the pair (f_{g_i}, τ_{g_i}) satisfies (6.3) and (6.4), $L(g_i, f_{g_i}, \tau_{g_i}) = 0$ and the implicit function argument from above implies that $P(g_i) = (f_{g_i}, \tau_{g_i})$ for large i. This proves the claim. In particular, we have shown that the constant c appearing above is equal to $-\nu_-(g)$. Since the map $g \mapsto (f_g, \tau_g)$ is analytic, the map

$$g \mapsto \nu_-(g) = -\tau_g(2\Delta f_g + |\nabla f_g|^2 - \mathrm{scal}_g) + f_g - n$$

is also analytic. This proves the lemma. □

Lemma 6.3.2. *Let (M, g_E) be a positive Einstein manifold. Then there exists a $C^{2,\alpha}$-neighbourhood $\mathcal{U}$ in the space of metrics and a constant $C > 0$ such that*

$$\left\| \frac{d}{dt}\Big|_{t=0} f_{g+th} \right\|_{C^{2,\alpha}} \leq C \|h\|_{C^{2,\alpha}}, \quad \left\| \frac{d}{dt}\Big|_{t=0} f_{g+th} \right\|_{H^i} \leq C \|h\|_{H^i}, \quad i = 0, 1, 2,$$
$$\left| \frac{d}{dt}\Big|_{t=0} \tau_{g+th} \right| \leq C \|h\|_{L^2}.$$

Proof. We obtain these estimates by deriving an Euler-Lagrange equation and using elliptic regularity. Recall that in a small neighbourhood of g_E, the pair (f_g, τ_g) realizing $\nu_-(g)$ is unique and satisfies the pair of equations

$$\tau(2\Delta f + |\nabla f|^2 - \mathrm{scal}) - f + n + \nu_- = 0, \tag{6.13}$$

$$\frac{1}{(4\pi\tau)^{n/2}} \int_M f e^{-f} \, dV = \frac{n}{2} + \nu_-. \tag{6.14}$$

Now we differentiate (6.13) and we obtain

$$\begin{aligned}\tau(2\dot{\Delta} f + 2\Delta\dot{f} - h(\mathrm{grad} f, \mathrm{grad} f) + 2\langle\nabla f, \nabla\dot{f}\rangle - \dot{\mathrm{scal}}) \\ + \dot{\tau}(2\Delta f + |\nabla f|^2 - \mathrm{scal}) - \dot{f} + \dot{\nu}_- = 0.\end{aligned} \tag{6.15}$$

Using (6.14), we can compute $\dot{\tau}$ in terms of f and $\dot{f}$. We have

$$\dot{\tau} = \frac{1}{4\pi}\frac{2}{n}\left(\frac{\int_M fe^{-f}\,dV}{\frac{n}{2}+\nu_-}\right)^{\frac{2}{n}-1}\left(\frac{\int_M fe^{-f}\,dV}{\frac{n}{2}+\nu_-}\right)^{\cdot} \tag{6.16}$$

and

$$\begin{aligned}\left(\frac{\int_M fe^{-f}\,dV}{\frac{n}{2}+\nu_-}\right)^{\cdot} =& \frac{\int_M (1-f)\dot{f}e^{-f}\,dV + \frac{1}{2}\int_M f\mathrm{tr}he^{-f}\,dV}{\frac{n}{2}+\nu_-} \\ &- \frac{\dot{\nu}_- \cdot \int_M fe^{-f}\,dV}{(\frac{n}{2}+\nu_-)^2}.\end{aligned} \tag{6.17}$$

Now we can seperate the terms of (6.15) which contain $\dot{f}$. Then we have

$$(2\tau\Delta - 1)\dot{f} + (2\Delta f + |\nabla f|^2 - \mathrm{scal})\int_M F\cdot\dot{f}\,dV + 2\tau\langle\nabla f, \nabla\dot{f}\rangle + (*) = 0,$$

where

$$\begin{aligned}F =& \frac{1}{2\pi n}\left(\frac{n}{2}+\nu_-\right)^{-2/n}\left(\int_M fe^{-f}\,dV\right)^{\frac{2}{n}-1}(1-f)e^{-f}, \\ (*) =& \tau(2\dot{\Delta} f - h(\mathrm{grad} f, \mathrm{grad} f) - \dot{\mathrm{scal}}) + \dot{\nu}_- \\ &+ \frac{1}{2\pi n}\left(\frac{\int_M fe^{-f}\,dV}{\frac{n}{2}+\nu_-}\right)^{\frac{2}{n}-1}\left(\frac{\frac{1}{2}\int_M f\mathrm{tr}he^{-f}\,dV}{\frac{n}{2}+\nu_-} - \frac{\dot{\nu}_-\cdot\int_M fe^{-f}\,dV}{(\frac{n}{2}+\nu_-)^2}\right).\end{aligned}$$

Now we define an integro-differential operator D by

$$Dv := (2\tau\Delta - 1)\,v + G\int_M F\cdot v\,dV + 2\tau\langle\nabla f, \nabla v\rangle, \tag{6.18}$$

where

$$G = 2\Delta f + |\nabla f|^2 - \mathrm{scal}.$$

Now, we can rewrite (6.15) as

$$D\dot{f} + (*) = 0. \tag{6.19}$$

On g_E, we have that $f = const$, $\tau = \frac{1}{2\mu}$, $F = \frac{1-f}{n\mu f\cdot\mathrm{vol}(M,g_E)}$ and $G = -n\mu$ so that D is equal to

$$D_{g_E}v = \left(\frac{1}{\mu}\Delta - 1\right)v - \frac{1-f}{f}\fint_M v\,dV_{g_E}.$$

Observe that this operator acts by multiplication with some nonzero constant on constant functions and as $(\frac{1}{\mu}\Delta - 1)$ on functions with vanishing integral. Therefore by Obata's eigenvalue estimate, D_{g_E} is an invertible operator and we have the estimates

$$\|v\|_{C^{2,\alpha}} \leq C_1 \|D_{g_E} v\|_{C^{0,\alpha}}, \qquad \|v\|_{H^i} \leq C_2 \|D_{g_E} v\|_{H^{i-2}}, \quad i = 0, 1, 2.$$

We show that these estimates also hold in a small $C^{2,\alpha}$-neighbourhood of g_E. We separate $F = F_0 + F_1$ and $G = G_0 + G_1$ where F_0, G_0 denote the constant parts (with respect to the underlying metric) of the functions F, G respectively. Then we have

$$\begin{aligned} Dv =& (2\tau\Delta - 1)v + G_0 \int_M F_0 \cdot v \, dV + G_1 \int_M F_0 \cdot v \, dV \\ &+ G_0 \int_M F_1 \cdot v \, dV + G_1 \int_M F_1 \cdot v \, dV + 2\tau \langle \nabla f, \nabla v \rangle. \end{aligned}$$

By Theorem 6.3.1, the mappings $g \mapsto F_0, F_1, G_0, G_1$ are smooth mappings from a $C^{2,\alpha}$-neighbourhood of g_E to $C^{2,\alpha}(M)$. Therefore,

$$D_0 v = (2\tau\Delta - 1)v + G_0 \int_M F_0 \cdot v \, dV$$

is also invertible for g close to g_E. For a fixed $\epsilon > 0$,

$$\begin{aligned} &\|v\|_{C^{2,\alpha}} \leq C_3 \left\| (2\tau\Delta - 1)v + G_0 \int_M F_0 \cdot v \, dV \right\|_{C^{0,\alpha}} \\ &\leq C_3 \|Dv\|_{C^{0,\alpha}} + C_3 \left\| G_0 \int_M F_1 \cdot v \, dV + G_1 \int_M F_1 \cdot v \, dV + 2\tau \langle \nabla f, \nabla v \rangle \right\|_{C^{0,\alpha}} \\ &\leq C_3 \|Dv\|_{C^{0,\alpha}} + \epsilon \|v\|_{C^{2,\alpha}} \end{aligned}$$

in a sufficiently small $C^{2,\alpha}$-neighbourhood of g_E, since the $C^{0,\alpha}$-norms of F_1, G_1 and ∇f are small there. Provided that we have chosen ϵ small enough, we obtain

$$\|v\|_{C^{2,\alpha}} \leq C_4 \|Dv\|_{C^{0,\alpha}},$$

and similarly,

$$\|v\|_{H^i} \leq C_5 \|Dv\|_{H^{i-2}}, \qquad i = 0, 1, 2,$$

in a small neighbourhood of g_E. Now, we have

$$\left\| \dot{f} \right\|_{C^{2,\alpha}} \leq C_6 \left\| D\dot{f} \right\|_{C^{0,\alpha}} \overset{(6.19)}{=} C_6 \|(*)\|_{C^{0,\alpha}} \leq C_7 \|h\|_{C^{2,\alpha}}.$$

The estimate of $(*)$ follows from the variational formulas for the Laplacian, the scalar curvature, the ν_--functional and the Hölder inequality. Analogously,

$$\left\| \dot{f} \right\|_{H^i} \leq C_8 \|h\|_{H^i}.$$

for $i = 0, 1, 2$. Finally, from (6.16) and (6.17),

$$|\dot{\tau}| \leq C_9 \left\| \dot{f} \right\|_{L^2} + C_{10} \|h\|_{L^2} \leq C_{11} \|h\|_{L^2},$$

which finishes the proof. □

Proposition 6.3.3 (Estimate of the second variation of ν_-). *Let (M, g_E) be a positive Einstein manifold. There exists a $C^{2,\alpha}$-neighbourhood $\mathcal{U}$ of g_E such that*

$$\left| \frac{d^2}{dsdt}\bigg|_{s,t=0} \nu_-(g+th+sk) \right| \leq C \left\| h \right\|_{H^1} \left\| k \right\|_{H^1}$$

for all $g \in \mathcal{U}$ and some constant $C > 0$.

Proof. We use similar arguments as in the proof of Proposition 5.3.3. Put $u = \frac{e^{-f}}{(4\pi\tau)^{n/2}}$ and $\nabla\nu_- = \tau(\mathrm{Ric} + \nabla^2 f) - \frac{1}{2}g$ so that the first variation of ν_- is

$$\nu_-(g)'(h) = -\int_M \langle \nabla\nu_-, h \rangle u \, dV.$$

As before, we use dot for t-derivatives and prime for s-derivatives. Then

$$\begin{aligned}
\frac{d^2}{dsdt}\bigg|_{s,t=0} \nu_-(g+th+sk) &= -\frac{d}{ds}\bigg|_{s=0} \int_M \langle \nabla\nu_-, h \rangle u \, dV \\
&= -\int_M \langle (\nabla\nu_-)', h \rangle u \, dV + 2\int_M \langle \nabla\nu_-, k \circ h \rangle u \, dV \\
&\quad - \int_M \langle \nabla\nu_-, h \rangle (u \, dV)'.
\end{aligned}$$

By standard estimates and Lemma 6.3.2, we even have

$$\begin{aligned}
\left| 2\int_M \langle \nabla\nu_-, k \circ h \rangle u \, dV \right| &\leq C_1 \left\| h \right\|_{L^2} \left\| k \right\|_{L^2}, \\
\left| \int_M \langle \nabla\nu_-, h \rangle (u \, dV)' \right| &\leq C_2 \left\| h \right\|_{L^2} \left\| k \right\|_{L^2}.
\end{aligned}$$

By the variational formula of the Ricci tensor and the Hessian and Lemma 6.3.2 again,

$$\left| \int_M \langle (\nabla\nu_-)', h \rangle u \, dV \right| \leq C_3 \left\| h \right\|_{H^1} \left\| k \right\|_{H^1},$$

which finishes the proof. □

Lemma 6.3.4. *Let (M, g_E) be a positive Einstein manifold. Then there exists a $C^{2,\alpha}$-neighbourhood $\mathcal{U}$ of g_E and a constant $C > 0$ such that*

$$\left\| \frac{d^2}{dtds}\bigg|_{t,s=0} f_{g+sk+th} \right\|_{H^i} \leq C \left\| h \right\|_{C^{2,\alpha}} \left\| k \right\|_{H^i}, \qquad i = 1, 2.$$

Proof. We again deal with the Euler-Lagrange equations satisfied by the pair (f_g, τ_g):

$$\tau(2\Delta f + |\nabla f|^2 - \mathrm{scal}) - f + n + \nu_- = 0, \tag{6.20}$$

$$\frac{1}{(4\pi\tau)^{n/2}} \int_M f e^{-f} \, dV = \frac{n}{2} + \nu_-. \tag{6.21}$$

Differentiating (6.20) twice yields

$$\begin{aligned}\tau(2\dot{\Delta}'f+2\dot{\Delta}f'+2\Delta'\dot{f}+2\Delta\dot{f}'-2h(\mathrm{grad}f,\mathrm{grad}f')\\ -2k(\mathrm{grad}f,\mathrm{grad}\dot{f})+2\langle\nabla f,\nabla\dot{f}'\rangle+2\langle\nabla\dot{f},\nabla f'\rangle-\dot{\mathrm{scal}}')\\ +\dot{\tau}(2\Delta'f+2\Delta f'+2\langle\nabla f,\nabla f'\rangle-k(\mathrm{grad}f,\mathrm{grad}f))\\ +\tau'(2\dot{\Delta}f+2\Delta\dot{f}+2\langle\nabla f,\nabla\dot{f}\rangle-h(\mathrm{grad}f,\mathrm{grad}f))\\ +\dot{\tau}'(2\Delta f+|\nabla f|^2-\mathrm{scal})-\dot{f}'+\dot{\nu}_-'=0.\end{aligned} \tag{6.22}$$

Using (6.21), we can compute $\dot{\tau}'$ in terms of $\dot{f}'$, $\dot{f}$ and f':

$$\begin{aligned}\dot{\tau}'=&\frac{1}{4\pi}\frac{2}{n}\left(\frac{2}{n}-1\right)\left(\frac{\int_M fe^{-f}\,dV}{\frac{n}{2}+\nu_-}\right)^{\frac{2}{n}-2}\left(\frac{\int_M fe^{-f}\,dV}{\frac{n}{2}+\nu_-}\right)^{\cdot}\left(\frac{\int_M fe^{-f}\,dV}{\frac{n}{2}+\nu_-}\right)'\\ &+\frac{1}{4\pi}\frac{2}{n}\left(\frac{\int_M fe^{-f}\,dV}{\frac{n}{2}+\nu_-}\right)^{\frac{2}{n}-1}\left(\frac{\int_M fe^{-f}\,dV}{\frac{n}{2}+\nu_-}\right)^{\cdot\,'}.\end{aligned}$$

We seperate the term containing $\dot{f}'$ and estimate all others. By Lemma 6.3.2 and the first variation of ν_-, the first of the two terms has an upper bound of the form $C\|h\|_{L^2}\|k\|_{L^2}$. Let us consider the second term more carefully. We have

$$\begin{aligned}\left(\frac{\int_M fe^{-f}\,dV}{\frac{n}{2}+\nu_-}\right)^{\cdot\,'}=&\frac{(\int_M fe^{-f}\,dV)^{\cdot\,'}}{\frac{n}{2}+\nu_-}+\frac{2\nu_-'\cdot\dot{\nu}_-\cdot\int_M fe^{-f}\,dV}{(\frac{n}{2}+\nu_-)^3}\\ &-\frac{\dot{\nu}_-(\int_M fe^{-f}\,dV)'+\nu_-'(\int_M fe^{-f}\,dV)^{\cdot}+\dot{\nu}_-'(\int_M fe^{-f}\,dV)'}{(\frac{n}{2}+\nu_-)^2}\end{aligned}$$

and

$$\begin{aligned}\left(\int_M fe^{-f}\,dV\right)^{\cdot\,'}=&\int_M[(1-f)\dot{f}'+(f-2)\dot{f}f'+\frac{1}{2}(1-f)\dot{f}\mathrm{tr}k]e^{-f}\,dV\\ &+\frac{1}{2}\int_M(1-f)f'\mathrm{tr}he^{-f}\,dV+\frac{1}{4}\int_M(\mathrm{tr}h\cdot\mathrm{tr}k-2\langle h,k\rangle)fe^{-f}\,dV.\end{aligned}$$

Thus,

$$\dot{\tau}'=\frac{1}{4\pi}\frac{2}{n}\left(\frac{\int_M fe^{-f}\,dV}{\frac{n}{2}+\nu_-}\right)^{\frac{2}{n}-1}\frac{\int_M(1-f)\dot{f}'e^{-f}\,dV}{\frac{n}{2}+\nu_-}+(A),$$

where (A) consists all terms which contain at most first derivatives of f. By Lemma 6.3.2, the first variational formula of ν_- and Proposition 6.3.3, we have the estimate

$$|(A)|\leq C\|h\|_{L^2}\|k\|_{L^2}. \tag{6.23}$$

Now we consider (6.22) again and separate the terms which contain $\dot{f}'$. Then we obtain

$$D\dot{f}'+(B)=0,$$

where D is the differential operator defined in (6.18) and

$$\begin{aligned}(B) =& \tau(2\dot{\Delta}'f - \dot{\text{scal}}') + \dot{\nu}_-' + (A)\cdot(2\Delta f + |\nabla f|^2 - \text{scal})\\ &+ \tau(2\Delta'\dot{f} + 2\dot{\Delta}f' - 2h(\text{grad}f, \text{grad}f') - 2k(\text{grad}f, \text{grad}\dot{f}) + 2\langle\nabla\dot{f}, \nabla f'\rangle)\\ &+ \dot{\tau}(2\Delta'f + 2\Delta f' + 2\langle\nabla f, \nabla f'\rangle - k(\text{grad}f, \text{grad}f))\\ &+ \tau'(2\dot{\Delta}f + 2\Delta\dot{f} + 2\langle\nabla f, \nabla\dot{f}\rangle - h(\text{grad}f, \text{grad}f)).\end{aligned}$$

In the proof of Lemma 6.3.2, we have shown that $D : H^i \to H^{i-2}$ is an isomorphism if we are in a small neighbourhood of g_E. From the first two variational formulas of the Laplacian and the scalar curvature, Lemma 6.3.2, Proposition 6.3.3, (6.23) and the Hölder inequality, we thus have

$$\left\|\dot{f}'\right\|_{H^i} \leq C\left\|D\dot{f}'\right\|_{H^{i-2}} = C\left\|(B)\right\|_{H^{i-2}} \leq C\left\|h\right\|_{C^{2,\alpha}}\left\|k\right\|_{H^i},$$

and the proof is finished. □

Proposition 6.3.5 (Estimates of the third variation of ν_-). *Let (M, g_E) be a positive Einstein manifold. There exists a $C^{2,\alpha}$-neighbourhood $\mathcal{U}$ of g_E such that*

$$\left|\left.\frac{d^3}{dt^3}\right|_{t=0}\nu_-(g+th)\right| \leq C\left\|h\right\|_{H^1}^2\left\|h\right\|_{C^{2,\alpha}}$$

for all $g \in \mathcal{U}$ and some constant $C > 0$.

Proof. We again put $u = \frac{e^{-f}}{(4\pi\tau)^{n/2}}$ and $\nabla\nu_- = \tau(\text{Ric} + \nabla^2 f) - \frac{1}{2}g$. Then

$$\begin{aligned}\left.\frac{d^3}{dt^3}\right|_{t=0}\nu_-(g+th) =& -\left.\frac{d^2}{dt^2}\right|_{t=0}\int_M\langle\nabla v_-, h\rangle u\, dV\\ =& -\int_M\langle(\nabla v_-)'', h\rangle u\, dV - 6\int_M\langle\nabla v_-, h\circ h\circ h\rangle u\, dV\\ &-\int_M\langle\nabla v_-, h\rangle(u\, dV)'' + 2\int_M\langle(\nabla v_-)', h\circ h\rangle u\, dV\\ &+2\int_M\langle\nabla v_-, h\circ h\rangle(u\, dV)' - \int_M\langle(\nabla v_-)', h\rangle(u\, dV)'.\end{aligned}$$

Further computations, standard estimates and the Lemmas 6.3.2 and 6.3.4 yield an upper bound of the form $C\left\|h\right\|_{H^1}^2\left\|h\right\|_{C^{2,\alpha}}$ for each of these terms (see also the proof of Proposition 5.3.5). □

6.4 The Integrable Case

As in Section 5.4, we prove stability/instability results under the assumption that all infinitesimal Einstein deformations are integrable. Additionally, we assume that 2μ (where μ is the Einstein constant) is not an eigenvalue of the Laplacian. These conditions are assumed to hold throughout this section.

6.4.1 Local Maximum of the Shrinker Entropy

In this subsection, we prove the analogue of Theorem 5.4.3 with the same methods. We use a similar notation to that in Subsection 5.4.1. Let $\mathcal{U}$ be a small $C^{2,\alpha}$-neighbourhood of a positive Einstein metric g_E and let

$$\mathcal{S}_{g_E} = \mathcal{U} \cap (g_E + \delta_{g_E}^{-1}(0))$$

be an affine slice of g_E in the space of metrics. Let

$$\mathcal{E} = \{g \in \mathcal{S}_{g_E} \mid \mathrm{Ric}_g = \alpha g \text{ for some } \alpha \in \mathbb{R}\}$$

be the set of Einstein metrics in the affine slice near g_E. Let

$$\mathcal{P} = \{g \in \mathcal{E} \mid \mathrm{Ric}_g = \mu g\} = \{g \in \mathcal{E} \mid \mathrm{vol}(M,g) = \mathrm{vol}(M,g_E)\},$$

where μ is the Einstein constant of g_E. If we assume that all infinitesimal Einstein deformations of g_E are integrable, $\mathcal{E}$ is a manifold near g_E and the tangent space at g_E is given by

$$T_{g_E}\mathcal{E} = \mathbb{R} \cdot g_E \oplus \ker(\Delta_E|_{TT}).$$

For any $g \in \mathcal{E}$,

$$\nu_-(g) = \log(\mathrm{vol}(M,g)) + \frac{n}{2}\log(\mathrm{scal}_g) + \frac{n}{2}(1 - \log(2\pi n)),$$

and thus, ν_- is constant on $\mathcal{P}$. By scale invariance, it is also constant on $\mathcal{E}$.

Let N be the L^2-orthogonal complement of $T_{g_E}\mathcal{E}$ in $\delta_{g_E}^{-1}(0)$. Then by the implicit function theorem, every $g \in \mathcal{S}_{g_E}$ can be written as $g = \bar{g} + h$, where $\bar{g} \in \mathcal{E}$ and $h \in N$. Since $\mathcal{P}$ (and hence also $\mathcal{E}$)) only contains smooth elements as was already discussed in Section 5.4.1, g is smooth if and only if h is smooth.

Theorem 6.4.1. *Let (M, g_E) be a positive Einstein manifold with constant μ. Suppose that g_E is Einstein-Hilbert stable and that the smallest nonzero eigenvalue of the Laplacian satisfies $\lambda > 2\mu$. Then there exists a small $C^{2,\alpha}$-neighbourhood $\mathcal{U} \subset \mathcal{M}$ of g_E such that $\nu_-(g) \leq \nu_-(g_E)$ for all $g \in \mathcal{U}$. Moreover, equality holds if and only if (M, g) is also Einstein.*

Proof. We first show that the second variation of ν_- vanishes on $T_{g_E}\mathcal{E}$ and is negative definite on N. The tangent space of the slice $\mathcal{S}_{g_E}$ splits as

$$\delta_{g_E}^{-1}(0) = \mathbb{R} \cdot g_E \oplus C_{g_E}(C^\infty_{g_E}(M)) \oplus TT_{g_E}$$

On $\mathbb{R} \cdot g_E$, the second variation vanishes whereas on $C_{g_E}(C^\infty_{g_E}(M)) \oplus TT_{g_E}$, it is defined by $-\frac{1}{4\mu \mathrm{vol}(M,g_E)}\Delta_E$. By the proof of Lemma 6.2.5, we have that $\Delta_E(C_{g_E}f) = C_{g_E}((\Delta - 2\mu)f)$ for $f \in C^\infty_{g_E}(M)$. The assumption on the spectrum of the Laplacian ensures that $-\frac{1}{4\mu \mathrm{vol}(M,g_E)}\Delta_E$ is negative on $C_{g_E}(C^\infty_{g_E}(M))$. By Einstein-Hilbert stability, the second variation is negative on TT-tensors orthogonal to $\ker\Delta_E|_{TT}$ and vanishes on $\ker\Delta_E|_{TT}$.

Now, we prove that g_E is a local maximum on $\mathcal{S}_{g_E}$ and the maximum is only attained on $\mathcal{E}$. By Taylor expansion,

$$\nu_-(\bar{g} + h) = \nu_-(\bar{g}) + \frac{1}{2}\frac{d^2}{dt^2}\bigg|_{t=0} \nu_-(\bar{g} + th) + R(\bar{g}, h),$$

$$R(\bar{g}, h) = \int_0^1 \left(\frac{1}{2} - t + \frac{1}{2}t^2\right) \frac{d^3}{dt^3}\nu_-(\bar{g} + th)dt,$$

where $\bar{g} \in \mathcal{E}$ and $h \in N$. As in the proof of Theorem 5.4.3, one shows that there are uniform bounds

$$\left.\frac{d^2}{dt^2}\right|_{t=0} \nu_-(\bar{g}+th) \leq -C_1 \|h\|_{H^1}^2,$$
$$R(\bar{g},h) \leq C_2 \|h\|_{C^{2,\alpha}} \|h\|_{H^1}^2.$$

Therefore, if we choose the $C^{2,\alpha}$-neighbourhood small enough, we have that $\nu_-(\bar{g}+h) \leq \nu_-(\bar{g}) = \nu_-(g_E)$ and equality holds if and only if $h = 0$. By the slice theorem, any metric $g \in \mathcal{U}$ can be written as $g = \varphi^*(\bar{g}+h)$ where $\varphi \in \mathrm{Diff}(M)$, $h \in N$ and $\bar{g} \in \mathcal{E}$. Thus,

$$\nu_-(g) = \nu_-(\bar{g}+h) \leq \nu_-(\bar{g}) = \nu_-(g_E),$$

and equality holds if and only if g is Einstein. □

6.4.2 A Lojasiewicz-Simon Inequality and Transversality

In this subsection, we prove analogoues of the results in Subsection 5.4.2.

Theorem 6.4.2 (Optimal Lojasiewicz-Simon inequality for ν_-). *Let (M, g_E) be a positive Einstein manifold with constant μ. Then there exists a $C^{2,\alpha}$-neighbourhood $\mathcal{U}$ of g_E and a constant $C > 0$ such that*

$$|\nu_-(g) - \nu_-(g_E)|^{1/2} \leq C \left\| \tau_g(\mathrm{Ric}_g + \nabla^2 f_g) - \frac{1}{2}g \right\|_{L^2}$$

for all $g \in \mathcal{U}$.

Theorem 6.4.3 (Transversality). *Let (M, g_E) be positive Einstein manifold with constant μ. Then there exists a $C^{2,\alpha}$-neighbourhood $\mathcal{U}$ of g_E and a constant $C > 0$ such that*

$$\left\| \mathrm{Ric}_g - \frac{1}{n}\left(⨍ \mathrm{scal}_g \, dV \right) g \right\|_{L^2} \leq C \left\| \tau_g(\mathrm{Ric}_g + \nabla^2 f_g) - \frac{1}{2}g \right\|_{L^2}$$

for all $g \in \mathcal{U}$.

Proof of Theorem 6.4.2 and Theorem 6.4.3. By diffeomorphism invariance, it suffices to prove these two inequalities on an affine slice in the space of metrics. Let $\mathcal{S}_{g_E}$, N and $\mathcal{E}$ be as above. Then every $g \in \mathcal{S}_{g_E}$ can be written as $g = \bar{g}+h$ where $\bar{g} \in \mathcal{E}$ and $h \in N$. By Taylor expansion and the Lemmas 6.3.2 and 6.3.4,

$$|\nu_-(\bar{g}+h) - \nu_-(\bar{g})| \leq C_1 \|h\|_{H^2}^2,$$
$$\left\| \mathrm{Ric}_{\bar{g}+h} - \frac{1}{n}\left(⨍ \mathrm{scal}_{\bar{g}+h} \, dV \right)(\bar{g}+h) \right\|_{L^2}^2 \leq C_2 \|h\|_{H^2}^2,$$

so it remains to show

$$\left\| \tau_{\bar{g}+h}(\mathrm{Ric}_{\bar{g}+h} + \nabla^2 f_{\bar{g}+h}) - \frac{1}{2}(\bar{g}+h) \right\|_{L^2} \geq C_3 \|h\|_{H^2}. \tag{6.24}$$

We put $\nabla\nu_-(g) = \tau_g(\mathrm{Ric} + \nabla^2 f_g) - \frac{1}{2}g$. Then by Lemma 5.4.7 and Lemma 6.2.3 (iii), we have

$$\nabla\nu_-(\bar{g}+h) = \frac{1}{4\mu}(\Delta_E)_{g_E}h + O_1 + O_2,$$

where

$$O_1 = \int_0^1 (1-t)\frac{d^2}{dt^2}\nabla\nu_-(g_E + th)dt,$$
$$O_2 = \int_0^1\int_0^1 \frac{d^2}{dsdt}\nabla\nu_-(g_E + s(g-g_E) + th)dtds.$$

By standard estimates and Lemmas 6.3.2 and 6.3.4,

$$\|O_1\|_{L^2} \leq C\|h\|_{C^{2,\alpha}}\|h\|_{H^2},$$
$$\|O_2\|_{L^2} \leq C\|\bar{g}-g_E\|_{C^{2,\alpha}}\|h\|_{H^2}.$$

By the eigenvalue assumption, $(\Delta_E)_{g_E}|_N$ is injective. Thus,

$$\begin{aligned}\|\nabla\nu_-(\bar{g}+h)\|_{L^2}^2 &= \frac{1}{16\mu^2}\|(\Delta_E)_{g_E}h\|_{L^2}^2 - \langle O_1+O_2, (\Delta_E)_{g_E}h\rangle + \|O_1+O_2\|_{L^2}^2 \\ &\geq C_1\|h\|_{H^2}^2 - C_2(\|O_1\|_{L^2} + \|O_2\|_{L^2})\|h\|_{H^2}.\end{aligned}$$

Therefore, if the neighbourhood is small enough, we obtain (6.24). □

6.4.3 Dynamical Stability and Instability

Lemma 6.4.4. *Let (M, g_E) be a positive Einstein manifold. For each $\epsilon > 0$ there exists $\delta > 0$ such that if $\|g_0 - g_E\|_{C^{k+2}} < \delta$, the Ricci flow (6.1) starting at g_0 exists on $[0,1]$ and satisfies*

$$\|g(t) - g_E\|_{C^k} < \epsilon$$

for all $t \in [0,1]$.

Proof. The Riemann curvature tensor and the Ricci tensor evolve under the standard Ricci flow as $\partial_t R = -\Delta R + R * R$, $\partial_t\mathrm{Ric} = -\Delta\mathrm{Ric} + R * \mathrm{Ric}$. Under the normalized Ricci flow, we have the evolution equations

$$\begin{aligned}\partial_t R &= -\Delta R + R * R + \frac{4}{n}\left(\fint_M \mathrm{scal}\, dV\right)R,\\ \partial_t\mathrm{Ric} &= -\Delta\mathrm{Ric} + R * \mathrm{Ric},\\ \partial_t\frac{1}{n}\left(\fint_M \mathrm{scal}\, dV\right)\cdot g &= \frac{2}{n}\left(\fint\left\langle \mathrm{Ric} - \frac{1}{n}\left(\fint_M \mathrm{scal}\, dV\right)\cdot g, G\right\rangle dV\right)\cdot g\\ &\quad - \frac{2}{n}\left(\fint_M \mathrm{scal}\, dV\right)\left(\mathrm{Ric} - \frac{1}{n}\left(\fint_M \mathrm{scal}\, dV\right)\cdot g\right),\end{aligned}$$

where G is the Einstein tensor. Let $\mathrm{Ric}^0 = \mathrm{Ric} - \frac{1}{n}\left(\fint_M \mathrm{scal}\, dV\right)g$. We then obtain the evolution inequalities

$$\partial_t |\nabla^i R|^2 \leq -\Delta |\nabla^i R|^2 + \sum_{j=1}^{i-1} C_{ij} |\nabla^j R| |\nabla^{i-j} R| |\nabla^i R| + C_{i1} (\sup_{p \in M} |R|) |\nabla^i R|^2,$$

$$\partial_t |\nabla^i \mathrm{Ric}^0|^2 \leq -\Delta |\nabla^i \mathrm{Ric}^0|^2 + \sum_{j=0}^{i} \tilde{C}_{ij} (\sup_{p \in M} |\nabla^j R| |\nabla^{i-j} \mathrm{Ric}|) |\nabla^i \mathrm{Ric}^0|.$$

Now one uses the maximum principle for scalars. The rest of the proof is exactly as in Lemma 5.4.10. □

Lemma 6.4.5. *Let $g(t)$, $t \in [0,T]$ be a solution of the Ricci flow (6.1) and suppose that*

$$\sup_{p \in M} |R_{g(t)}|_{g(t)} \leq T^{-1} \qquad \forall t \in [0,T].$$

Then for each $k \geq 1$, there exists a constant $C(k)$ such that

$$\sup_{p \in M} |\nabla^k R_{g(t)}|_{g(t)} \leq C(k) \cdot T^{-1} t^{-k/2} \qquad \forall t \in (0,T].$$

Proof. By the evolution equation $\partial_t R = -\Delta R + R * R + \frac{4}{n}(\fint_M \mathrm{scal}\, dV) R$, we have the evolution inequality

$$\begin{aligned}\partial_t |\nabla^i R|^2 \leq & -\Delta |\nabla^i R|^2 - 2|\nabla^{i+1} R|^2 + \sum_{j=1}^{i-1} C_{ij} |\nabla^j R| |\nabla^{i-j} R| |\nabla^i R| \\ & + C_{i1} (\sup_{p \in M} |R|) |\nabla^i R|^2.\end{aligned}$$

The lemma is shown by induction on k. This works exactly as in the proof of Lemma 5.4.11. □

Remark 6.4.6. As in Remark 5.4.12, we obtain uniform bounds of all derivatives of the curvature along the Ricci flow on $[\delta, T]$, if the curvature is bounded on $[0,T]$.

Theorem 6.4.7 (Dynamical stability). *Let (M, g_E) be a compact positive Einstein manifold with constant μ which is Einstein-Hilbert stable. Suppose that the integrability condition holds and that the smallest nonzero eigenvalue of the Laplacian satisfies $\lambda > 2\mu$. Let $k \geq 3$.*

Then for every C^k-neighbourhood $\mathcal{U}$ of g_E in the space of metrics, there exists a C^{k+2}-neighbourhood $\mathcal{V}$ such that the Ricci flow, starting at any $g_0 \in \mathcal{V}$, stays in $\mathcal{U}$ for all time and converges to an Einstein metric $g_\infty \in \mathcal{U}$. The convergence is exponentially, i.e. there exist constants $C_1, C_2 > 0$ such that for all $t \geq 0$,

$$\|g(t) - g_E\|_{C^k_{g_E}} \leq C_1 e^{-C_2 t}.$$

Proof. As above, we denote by $\mathcal{B}^k_\epsilon$ the ϵ-ball around g_E with respect to the $C^k_{g_E}$-norm. Without loss of generality, we assume that $\mathcal{U} = \mathcal{B}^k_\epsilon$ for an $\epsilon > 0$ so small that Theorems 6.4.1, 6.4.2 and 6.4.3 hold on $\mathcal{U}$. By Lemma 6.4.4, we can choose $\mathcal{V}$ so small that any Ricci flow starting in $\mathcal{V}$ stays in $\mathcal{B}^k_{\epsilon/4}$ up to time 1.

Let now $T \geq 1$ be the maximal time such that any Ricci flow starting in $\mathcal{V}$ stays in $\mathcal{U}$ for all $t < T$. By definition of T, we have uniform curvature bounds

$$\sup_{p \in M} |R_{g(t)}|_{g(t)} \leq C_1 \qquad \forall t \in [0,T),$$

and by Remark 6.4.6,

$$\sup_{p \in M} |\nabla^i R_{g(t)}|_{g(t)} \leq C(i) \qquad \forall t \in [1,T),\ \forall i \geq 0.$$

Assume that $\epsilon > 0$ is so small that the C^k-norms defined by g_E and $g(t)$ differ at most by a factor 2. Then we have

$$\begin{aligned}\|g(T) - g_E\|_{C^k_{g_E}} &\leq \|g(1) - g_E\|_{C^k_{g_E}} + \int_1^T \frac{d}{dt} \|g(t) - g(1)\|_{C^k_{g_E}} dt \\ &\leq \frac{\epsilon}{4} + 4 \int_1^T \left\| \mathrm{Ric}^0_{g(t)} \right\|_{C^k_{g(t)}} dt.\end{aligned}$$

By interpolation (c.f. [Ham82, Corollary 12.7]), using the bounds on $|\nabla^i R|$,

$$\left\| \mathrm{Ric}^0_{g(t)} \right\|_{C^k_{g(t)}} \leq C_2 \left\| \mathrm{Ric}^0_{g(t)} \right\|_{H^l} \leq C_3 \left\| \mathrm{Ric}^0_{g(t)} \right\|^{\beta}_{L^2}$$

for some $\beta \in (0,1)$ and $C > 0$. Here, $l > k$ is some constant such that Sobolev embedding holds. By Theorems 6.4.1, 6.4.2 and 6.4.3,

$$\begin{aligned}-\frac{d}{dt}|\nu_-(g(t)) - \nu_-(g_E)|^{\beta/2} &= \frac{\beta}{2}|\nu_-(g(t)) - \nu_-(g_E)|^{\beta/2-1} \frac{d}{dt}\nu_-(g(t)) \\ &\geq C_4 |\nu_-(g(t)) - \nu_-(g_E)|^{\beta/2-1} \left\| \nabla (\nu_-)_{g(t)} \right\|^2_{L^2} \\ &\geq C_5 \left\| \mathrm{Ric}^0_{g(t)} \right\|^{\beta}_{L^2} \geq C_6 \left\| \mathrm{Ric}^0_{g(t)} \right\|_{C^k_{g(t)}}.\end{aligned}$$

Hence by integration,

$$\begin{aligned}\int_1^T \left\| \mathrm{Ric}^0_{g(t)} \right\|_{C^k_{g(t)}} dt &\leq C_7 |\nu_-(g(1)) - \nu_-(g_E)|^{\beta/2} \\ &\leq C_7 |\nu_-(g(0)) - \nu_-(g_E)|^{\beta/2} \leq \frac{\epsilon}{16},\end{aligned}$$

provided that we have chosen $\mathcal{V}$ small enough. This shows that $T = \infty$. Since $\int_M \|\dot{g}(t)\|_{C^k_{g_E}} dt < \infty$, $g(t)$ converges to some limit g_∞ as $t \to \infty$. By Theorem 6.4.2, we have $-\frac{d}{dt}|\nu_-(g(t)) - \nu_-(g_E)| \geq C_8 |\nu_-(g(t)) - \nu_-(g_E)|$. Thus,

$$|\nu_-(g(t)) - \nu_-(g_E)| \leq e^{C_8 t} |\nu_-(g_0) - \nu_-(g_E)|,$$

which shows that $\nu_-(g_\infty) = \nu_-(g_E)$ and by Theorem 6.4.1, g_∞ is Einstein. The convergence is exponential, since for $t_1 < t_2$,

$$\begin{aligned}\|g(t_1) - g(t_2)\|_{C^k_{g_E}} &\leq C_9 |\nu_-(g(t_1)) - \nu_-(g_E)|^{\beta/2} \\ &\leq C_9 e^{-\frac{C_8 \beta}{2} t_1} |\nu_-(g_0) - \nu_-(g_E)|^{\beta/2}.\end{aligned}$$

The assertion follows from $t_2 \to \infty$. $\square$

Theorem 6.4.8 (Dynamical instability). *Let (M, g_E) be a positive Einstein manifold with constant μ which satisfies the integrability condition. Suppose that $2\mu \notin \mathrm{spec}(\Delta)$. If (M, g_E) is Einstein-Hilbert unstable or we have that $(\frac{n}{n-1}\mu, 2\mu) \cap \mathrm{spec}(\Delta) \neq \emptyset$, there exists a nontrivial ancient Ricci flow emerging from it, i.e. there is a Ricci flow $g(t)$, defined on $t \in (-\infty, T]$, such that $\lim_{t\to-\infty} g(t) = g_E$.*

Proof. Under these conditions, (M, g_E) cannot be a local maximum of ν_-. Let $g_i \to g_E$ in C^k and suppose that $\nu_-(g_i) > \nu_-(g_E)$ for all i. Let $g_i(t)$ be the Ricci flow (6.2) starting at g_i. Then by Lemma 6.4.4, $\bar{g}_i = g_i(1)$ converges to g_E in C^{k-2} and by monotonicity, $\nu_-(\bar{g}_i) > \nu_-(g_E)$ as well. Let $\epsilon > 0$ be so small that Theorems 6.4.2 and 6.4.3 both hold on $\mathcal{B}^{k-2}_{2\epsilon}$. Theorem 6.4.2 yields the differential inequality

$$\frac{d}{dt}(\nu_-(g_i(t)) - \nu_-(g_E)) \geq C_1(\nu_-(g_i(t)) - \nu_-(g_E)),$$

from which we obtain

$$(\nu_-(g_i(t)) - \nu_-(g_E))e^{C_1(s-t)} \leq (\nu_-(g_i(s)) - \nu_-(g_E)), \tag{6.25}$$

as long as g_i stays in $\mathcal{B}^{k-2}_{2\epsilon}$. Thus, there exists a t_i such that

$$\|g_i(t_i) - g_E\|_{C^{k-2}} = \epsilon,$$

and $t_i \to \infty$. If t_i was bounded, $g_i(t_i) \to g_E$ in C^{k-2}. By interpolation,

$$\left\|\mathrm{Ric}^0_{g_i(t)}\right\|_{C^{k-2}} \leq C_2 \left\|\mathrm{Ric}^0_{g_i(t)}\right\|^{\beta}_{L^2} \tag{6.26}$$

for some $\beta \in (0,1)$. By Theorems 6.4.2 and 6.4.3, we have the differential inequality

$$\frac{d}{dt}(\nu_-(g_i(t)) - \nu_-(g_E))^{\beta/2} \geq C_3 \left\|\mathrm{Ric}^0_{g_i(t)}\right\|^{\beta}_{L^2}, \tag{6.27}$$

if $\nu_-(g_i(t)) > \nu_-(g_E)$. Thus by the triangle inequality and by integration,

$$\epsilon = \|g_i(t_i) - g_E\|_{C^{k-2}} \leq \|\bar{g}_i - g_E\|_{C^{k-2}} + C_4(\nu_-(g_i(t_i)) - \nu_-(g_E))^{\beta/2}. \tag{6.28}$$

Now, put $g_i^s(t) := g_i(t + t_i)$, $t \in [T_i, 0]$, where $T_i = 1 - t_i \to -\infty$. We have

$$\|g_i^s(t) - g_E\|_{C^{k-2}} \leq \epsilon \qquad \forall t \in [T_i, 0],$$
$$g_i^s(T_i) \to g_E \text{ in } C^{k-2}.$$

Because the embedding $C^{k-3}(M) \subset C^{k-2}(M)$ is compact, we can choose a subsequence of the g_i^s, converging in $C^{k-3}_{loc}(M \times (-\infty, 0])$ to an ancient Ricci flow $g(t)$, $t \in (-\infty, 0]$. From taking the limit $i \to \infty$ in (6.28), we have that $\epsilon \leq C_4(\nu_-(g(0)) - \nu_-(g_E))^{\beta/2}$ which shows that the Ricci flow is nontrivial. For $T_i \leq t$, we have, by (6.26) and (6.27),

$$\begin{aligned}\|g_i^s(T_i) - g_i^s(t)\|_{C^{k-3}} \leq& C_5(\nu_-(g_i(t + t_i)) - \nu_-(g_E))^{\beta/2}\\ \leq& C_5(\nu_-(g_i(t_i)) - \nu_-(g_E))^{\beta/2}e^{C_1 t} = C_6 e^{C_1 t}.\end{aligned}$$

Thus,

$$\|g_E - g(t)\|_{C^{k-3}} \leq \|g_E - g_i^s(T_i)\|_{C^{k-3}} + C_6 e^{C_1 t} + \|g_i^s(t) - g(t)\|_{C^{k-3}}.$$

It follows that $\|g_E - g(t)\|_{C^{k-3}} \to 0$ as $t \to -\infty$. □

Remark 6.4.9. In contrast to the negative case, many examples satisfying the assumptions of Theorem 6.4.8 are known. We already discussed some Einstein-Hilbert unstable examples (see e.g. Example 3.1.7). In fact, there are also examples of Einstein manifolds which are Einstein-Hilbert stable but do not satisfy the eigenvalue assumptions from above, e.g. $\mathbb{H}P^n$ for $n \geq 3$ (c.f. [CH13]).

Remark 6.4.10. The condition on the Laplacian spectrum appearing in Theorem 6.4.7 also plays a role in other contexts. Let (M, g) be Einstein with constant $\mu > 0$. The identity map on (M, g) is stable as a harmonic map if and only if the smallest nonzero eigenvalue of the Laplacian on an Einstein manifold (M, g) satisfies $\lambda \geq 2\mu$ (see [Smi75, Proposition 2.11]). The same condition on the Laplacian spectrum also ensures that simply-connected irreducible symmetric spaces of compact type are stable with respect to the functional $g \mapsto \int_M |R|^{n/2}\, dV$ restricted to its conformal class (c.f. [BM12, pp. 1-2]).

Recall also that this condition appeared when we discussed the spectrum of the Einstein operator on product spaces, c.f. Proposition 3.3.7.

6.5 The Nonintegrable Case

As in the negative case, we are also able to get rid of the integrability condition here. We prove analogues of Theorems 6.4.1 and 6.4.2. The proofs of these theorems are very similar to the proofs of Section 5.5.

6.5.1 Local Maximum of the Shrinker Entropy

Theorem 6.5.1. *Let (M, g_E) be a positive Einstein manifold with constant μ. If g_E is a local maximum of ν_-, it is a local maximum of the Yamabe functional and the smallest nonzero eigenvalue satisfies $\lambda \geq 2\mu$. Conversely, if g_E is a local maximum of the Yamabe functional and $\lambda > 2\mu$, then g_E is a local maximum of ν_-. In this case, any other local maximum is also an Einstein metric.*

Proof. Let $c = \mathrm{vol}(M, g_E)$ and recall the notations

$$\begin{aligned}\mathcal{C} &= \{g \in \mathcal{M} | \mathrm{scal}_g \text{ is constant}\},\\ \mathcal{C}_c &= \{g \in \mathcal{M} | \mathrm{scal}_g \text{ is constant and } \mathrm{vol}(M, g) = c\}.\end{aligned}$$

Since we excluded the case of the sphere, Obata's eigenvalue estimate implies that $\frac{\mathrm{scal}_{g_E}}{n-1} \notin \mathrm{spec}(\Delta_{g_E})$. Thus, the map

$$\begin{aligned}\Phi \colon C^\infty(M) \times \mathcal{C}_c &\to \mathcal{M},\\ (v, g) &\mapsto v \cdot g\end{aligned}$$

is a local ILH-diffeomorphism around $(1, g_E)$. We first evaluate ν_- on the space of constant scalar curvature metrics. Let $\bar{g}$ be a metric of constant scalar curvature and consider the pair

$$\bar{f} = \log(\mathrm{vol}(M, \bar{g})) + \frac{n}{2}\log(\mathrm{scal}_{\bar{g}}) - \frac{n}{2}\log(2\pi n), \qquad \bar{\tau} = \frac{n}{2\mathrm{scal}_{\bar{g}}}.$$

This pair satisfies the coupled Euler-Lagrange equations (6.3) and (6.4) and the constraint in the definition of ν_-. If $\bar{g}$ is close to g_E (in $C^{2,\alpha}$), the pair $(\bar{f}, \bar{\tau})$

is close to (f_{g_E}, τ_{g_E}). Therefore, by the implicit function argument used in the proof of Lemma 6.3.1, $(\bar{f}, \bar{\tau})$ is the pair realizing $\nu_-(\bar{g})$, provided that $\bar{g}$ is close enough to g_E. In other words, $(\bar{f}, \bar{\tau}) = (\tau_{\bar{g}}, f_{\bar{g}})$. In particular, $f_{\bar{g}}$ is constant. Thus,

$$\nu_-(\bar{g}) = \log(\mathrm{vol}(M, \bar{g})) + \frac{n}{2}\log(\mathrm{scal}_{\bar{g}}) + \frac{n}{2}(1 - \log(2\pi n)).$$

By the monotonicity of the logarithm and scale invariance of ν_-, g_E is a local maximum of the ν_- restricted to $\mathcal{C}$ if and only if g_E is a local maximum of the Einstein-Hilbert functional restricted to $\mathcal{C}_c$. Since all constant scalar curvature metrics in a sufficiently small neighbourhood of g_E are Yamabe metrics, this is equivalent to the assertion that g_E is a local maximum of the Yamabe functional. If g_E is a local maximum of ν_- on all of $\mathcal{M}$, the eigenvalue bound follows from Corollary 6.2.5.

We now investigate the behavior in conformal directions and use the eigenvalue assumption. Let $\bar{g}$ be of constant scalar curvature and $h = v\bar{g}$ for some $v \in C^\infty_{\bar{g}}(M)$. Then

$$\begin{aligned}
\nu'_-(\bar{g})(h) &= -\frac{1}{4\pi\tau_{\bar{g}}}\int_M \left\langle \tau_{\bar{g}}(\mathrm{Ric}_{\bar{g}} + \nabla^2 f_{\bar{g}}) - \frac{1}{2}\bar{g}, h\right\rangle e^{-f_{\bar{g}}}\, dV_{\bar{g}} \\
&= -⨍_M \left\langle \frac{n}{2\mathrm{scal}_{\bar{g}}}\mathrm{Ric}_{\bar{g}} - \frac{1}{2}\bar{g}, v\bar{g}\right\rangle\, dV_{\bar{g}} \\
&= -⨍_M \left(\frac{n}{2} - \frac{n}{2}\right) v\, dV_{\bar{g}} = 0.
\end{aligned}$$

The second variation is equal to

$$\begin{aligned}
&\left.\frac{d^2}{dt^2}\right|_{t=0} \nu_-(\bar{g} + th) \\
&= -\frac{1}{4\pi\tau_{\bar{g}}}\int_M \left\langle \left.\frac{d}{dt}\right|_{t=0} \tau_{\bar{g}+th}(\mathrm{Ric}_{\bar{g}+th} + \nabla^2 f_{\bar{g}+th}) - \frac{1}{2}h, h\right\rangle e^{-f_{\bar{g}}}\, dV_{\bar{g}} \\
&= -⨍_M \tau' \cdot v \cdot \mathrm{scal}_{\bar{g}}\, dV - ⨍_M \langle \tau(\mathrm{Ric}' + \nabla^2(f')) - \frac{1}{2}v\bar{g}, v\bar{g}\rangle\, dV_{\bar{g}} \\
&= -⨍_M \langle \tau\left(\frac{1}{2}\Delta_L(v\bar{g}) - \delta^*\delta(v\bar{g}) - \frac{n}{2}\nabla^2 v\right) + \nabla^2(f')) - \frac{1}{2}v\bar{g}, v\bar{g}\rangle\, dV_{\bar{g}} \\
&= -⨍_M \left\langle \tau\left(\frac{1}{2}(\Delta v)\bar{g} + \left(1 - \frac{n}{2}\right)\nabla^2 v + \nabla^2(f')\right) - \frac{1}{2}v\bar{g}, v\bar{g}\right\rangle\, dV_{\bar{g}} \\
&= -⨍_M \frac{n}{2}\left(\frac{n}{2\mathrm{scal}_{\bar{g}}}\Delta v - v\right) v\, dV - ⨍_M \tau\left\langle \left(1 - \frac{n}{2}\right)\nabla^2 v + \nabla^2(f'), v\bar{g}\right\rangle\, dV_{\bar{g}}.
\end{aligned}$$

We first deal with the terms containing the Hessians of v and f'. Differentiating the Euler-Lagrange equation

$$\tau(2\Delta f + |\nabla f|^2 - \mathrm{scal}) - g + n - \nu_- = 0$$

in the direction of $h = vg$ yields

$$\begin{aligned}
(2\tau\Delta - 1)f' &= \tau'\mathrm{scal}_{\bar{g}} + \tau\mathrm{scal}'_{\bar{g}} \\
&= \tau'\mathrm{scal}_{\bar{g}} + \tau((n-1)\Delta v - \mathrm{scal}_{\bar{g}} v).
\end{aligned}$$

Here we use the fact that $f_{\bar{g}}$ is constant since $\bar{g}$ is of constant scalar curvature. Since $\tau_{\bar{g}} = \frac{n}{2\mathrm{scal}_{\bar{g}}}$,

$$f' = \left(\frac{n}{\mathrm{scal}}\Delta - 1\right)^{-1}\left(\tau'\mathrm{scal}_{\bar{g}} + \frac{n}{2}\left(\frac{n-1}{\mathrm{scal}}\Delta - 1\right)v\right).$$

Because of the assumption on the spectrum of g_E, we were allowed to take the inverse of $\frac{n}{\mathrm{scal}}\Delta - 1$ for $\bar{g}$ close enough to g_E. Moreover,

$$\Delta f' = \frac{n}{2}\left(\frac{n}{\mathrm{scal}}\Delta - 1\right)^{-1}\left(\frac{n-1}{\mathrm{scal}}\Delta - 1\right)\Delta v.$$

Thus,

$$\begin{aligned}
&-⨍_M \tau\left\langle\left(1-\frac{n}{2}\right)\nabla^2 v + \nabla^2(f'), v\bar{g}\right\rangle\, dV \\
=&\frac{n}{2\mathrm{scal}}⨍_M\left(\left(1-\frac{n}{2}\right)\Delta v + \Delta f'\right)v\, dV \\
=&\frac{n}{2\mathrm{scal}}⨍_M\left((1-\frac{n}{2})\Delta v + \frac{n}{2}(\frac{n}{\mathrm{scal}}\Delta - 1)^{-1}\left(\frac{n-1}{\mathrm{scal}}\Delta - 1\right)\Delta v\right)v\, dV.
\end{aligned}$$

Therefore, the second variation is equal to

$$\frac{d^2}{dt^2}|_{t=0}\nu_-(\bar{g}+th) = -⨍_M Lv\cdot v\, dV, \tag{6.29}$$

where L is the linear operator given by

$$\begin{aligned}
L =&\frac{n}{2}\left(\frac{n}{2\mathrm{scal}_{\bar{g}}}\Delta - 1\right) \\
&-\frac{n}{2\mathrm{scal}}\left(\left(1-\frac{n}{2}\right)\Delta + \frac{n}{2}\left(\frac{n}{\mathrm{scal}}\Delta - 1\right)^{-1}\left(\frac{n-1}{\mathrm{scal}}\Delta - 1\right)\Delta\right) \\
=&\frac{n+1}{4}\left(\frac{n}{\mathrm{scal}}\Delta - 1\right)^{-1}\left(\frac{n}{\mathrm{scal}}\Delta - 2\right)\left(\frac{n}{\mathrm{scal}}\Delta - \frac{n}{n-1}\right).
\end{aligned}$$

If the smallest nonzero eigenvalue of the Laplacian is greater than $\frac{2\mathrm{scal}}{n}$, the operator $L\colon C^\infty_{\bar{g}}(M) \to C^\infty_{\bar{g}}(M)$ is positive. By assumption, this is certainly true in a small neighbourhood of g_E in the space of constant scalar curvature metrics. By continuity, if $\epsilon > 0$ is sufficiently small,

$$\begin{aligned}
-⨍_M Lv\cdot v\, dV &= -\epsilon ⨍_M |\nabla v|^2\, dV - ⨍_M (L-\epsilon\Delta)v\cdot v\, dV \\
&\leq -\epsilon' \|\nabla v\|^2_{L^2} - C_1\|v\|^2_{L^2} \\
&\leq -C_2\|v\|^2_{H^1},
\end{aligned}$$

and this estimate is uniformly in a small neighbourhood. Let now $g \in \mathcal{M}$ be an arbitrary metric in a small $C^{2,\alpha}$-neighbourhood of g_E. By the above, it can be written as $g = \tilde{v}\cdot\tilde{g}$, where $(\tilde{v},\tilde{g}) \in C^\infty(M)\times\mathcal{C}_{g_E}$ is close to $(1, g_E)$. By substituting

$$v = \frac{\tilde{v} - ⨍\tilde{v}\, dV_{\tilde{g}}}{⨍\tilde{v}\, dV_{\tilde{g}}}, \qquad \bar{g} = \left(⨍\tilde{v}\, dV_{\tilde{g}}\right)\tilde{g},$$

we can write $g = (1+v)\bar{g}$ where $\bar{g} \in \mathcal{C}$ is close to g_E, and $v \in C^\infty_{\bar{g}}(M)$ is close to 0. By Taylor expansion and Proposition 6.3.5,

$$\begin{aligned}\nu_-(g) &= \nu_-(\bar{g}) + \frac{1}{2}\frac{d^2}{dt^2}\bigg|_{t=0}\nu_-(\bar{g}+tv\bar{g}) + \int_0^1\left(\frac{1}{2}-t+\frac{1}{2}t^2\right)\frac{d^3}{dt^3}\nu_-(\bar{g}+tv\bar{g})dt\\ &\leq \nu_-(g_E) - \frac{C_2}{2}\left\|v\right\|^2_{H^1} + C_3\left\|v\right\|_{C^{2,\alpha}}\left\|v\right\|^2_{H^1}\\ &\leq \nu_-(g_E),\end{aligned}$$

provided that the neighbourhood is small enough. If $g = (1+v)\bar{g}$ is another local maximum of ν_-, then $v = 0$ and $g \in \mathcal{C}$ is a local maximum of the total scalar curvature restricted to $\mathcal{C}_d$. Here, $d = \text{vol}(M, g)$. By Proposition 2.6.2, g is Einstein. □

Remark 6.5.2. If (M, g_E) is a local maximum of the Yamabe functional and we have the weak inequality $\lambda \geq 2\mu$, then it is in general not true that it is a local maximum of ν_-. A counterexample will be given in Section 6.6.

Corollary 6.5.3. *Let (M, g_E) be a compact positive Einstein manifold with constant μ. If g_E is a local maximum of the Yamabe invariant and $\lambda > 2\mu$, any shrinking gradient Ricci soliton in a sufficiently small neighbourhood of g_E is nessecarily Einstein.*

Proof. This follows from Theorem 6.5.1 and the fact that shrinking gradient Ricci solitons are precisely the critical points of ν_-. □

6.5.2 A Lojasiewicz-Simon Inequality

Theorem 6.5.4 (Lojasiewicz-Simon inequality). *Let (M, g_E) be a positive Einstein manifold. Then there exists a $C^{2,\alpha}$ neighbourhood $\mathcal{U}$ of g_E and constants $\sigma \in [1/2, 1)$, $C > 0$ such that*

$$|\nu_-(g) - \nu_-(g_E)|^\sigma \leq C\left\|\tau(\text{Ric}_g + \nabla^2 f_g) - \frac{1}{2}g\right\|_{L^2} \tag{6.30}$$

for all $g \in \mathcal{U}$.

Proof. Since both sides are diffeomorphism invariant, it suffices to show the inequality on a slice to the action of the diffeomorphism group. Let

$$\mathcal{S}_{g_E} = \mathcal{U} \cap \left\{g_E + h \mid h \in \delta^{-1}_{g_E}(0)\right\}.$$

Let $\tilde{\nu}_-$ be the ν_--functional restricted to $\mathcal{S}_{g_E}$. Obviously, $\tilde{\nu}_-$ is analytic since ν_- is. By the first variational formula in Lemma 6.2.2, the L^2-gradient of ν_- is (up to a constant factor) given by $\nabla\nu_-(g) = [\tau(\text{Ric}_g + \nabla^2 f_g) - \frac{1}{2}g]e^{-f_g}$. It vanishes at g_E. On the neighbourhood $\mathcal{U}$, we have the uniform estimate

$$\left\|\nabla\nu_-(g_1) - \nabla\nu_-(g_2)\right\|_{L^2} \leq C\left\|g_1 - g_2\right\|_{H^2}, \tag{6.31}$$

which holds by Taylor expansion. The L^2-gradient of $\tilde{\nu}_-$ is given by the projection of $\nabla\nu_-$ to $\delta^{-1}_{g_E}(0)$. Therefore, (6.31) also holds for $\nabla\tilde{\nu}_-$. The linearization of $\tilde{\nu}_-$ at g_E vanishes on $\mathbb{R}\cdot g_E$ and equals $-\frac{1}{4\mu\text{vol}(M,g_E)}\Delta_E$ on the L^2-orthogonal complement of $\mathbb{R}\cdot g_E$ in $\delta_{g_E}(0)$, see Proposition 6.2.4. Let us denote this operator by D. By ellipticity,

$$D : (\delta_{g_E}^{-1}(0))^{C^{2,\alpha}} \to (\delta_{g_E}^{-1}(0))^{C^{0,\alpha}}$$

is Fredholm. It also satisfies the estimate $\|Dh\|_{L^2} \leq C \|h\|_{H^2}$. By Theorem [CM12, Theorem 6.3], there exists a constant $\sigma \in [1/2, 1)$ such that the inequality $|\nu_-(g) - \nu_-(g_E)|^\sigma \leq \|\nabla\tilde{\nu}_-(g)\|_{L^2}$ holds for any $g \in \mathcal{S}_{g_E}$. Since

$$\|\nabla\tilde{\nu}_-(g)\|_{L^2} \leq \|\nabla\nu_-(g)\|_{L^2} \leq C \left\|\tau(\mathrm{Ric}_g + \nabla^2 f_g) - \frac{1}{2}g\right\|_{L^2},$$

(5.25) holds on all $g \in \mathcal{S}_{g_E}$. By diffeomorphism invariance, it holds on all $g \in \mathcal{U}$. □

6.5.3 Dynamical Stability and Instability

In order to consider dynamical stability in the nonintegrable case, we have to deal with another variant of the Ricci flow, which is given by the differential equation

$$\dot{g}(t) = -2\mathrm{Ric}_{g(t)} + \frac{1}{\tau_{g(t)}}g(t). \tag{6.32}$$

This can be considered as the gradient flow of ν_- on the space of metrics modulo diffeomorphism. Suppose we have a solution $g(t)$ of $\dot{g}(t) = -2\mathrm{Ric}_{g(t)}$, then a solution $\tilde{g}(t)$ of (6.32) is given by

$$\tilde{g}(t) = v(t)^{-1} g\left(\int_0^t v(t')dt'\right),$$

where $v : [0, T) \to \mathbb{R}$ is some positive function satisfying the integro-differential equation

$$\dot{v}(t) = -v^2(t)\left(\tau_{g(\int_0^t v(t')dt')}\right)^{-1}$$

with initial condition $v(0) = 1$. In this subsection, we prove dynamical stability/instability results with respect to (6.32).

Lemma 6.5.5. *Let (M, g_E) be a positive Einstein manifold. For each $\epsilon > 0$ there exists $\delta > 0$ such that if $\|g_0 - g_E\|_{C^{k+2}} < \delta$, the Ricci flow (6.32) starting at g_0 exists on $[0, 1]$ and satisfies*

$$\|g(t) - g_E\|_{C^k} < \epsilon$$

for all $t \in [0, 1]$.

Proof. From the well-known evolution equations $\partial_t R = -\Delta R + R * R$ and $\partial_t \mathrm{Ric} = -\Delta\mathrm{Ric} + R * \mathrm{Ric}$ for the standard Ricci flow, we derive the evolution equations

$$\begin{aligned}
\partial_t R &= -\Delta R + R * R + \frac{2}{\tau}R, \\
\partial_t \mathrm{Ric} &= -\Delta\mathrm{Ric} + R * \mathrm{Ric}, \\
\partial_t \frac{1}{2\tau}g &= -\frac{\partial_t\tau}{2\tau^2}g + \frac{1}{2\tau}\left(-2\mathrm{Ric} + \frac{1}{\tau}g\right)
\end{aligned}$$

for the flow (6.32). From these, we obtain the evolution inequality

$$\partial_t|\nabla^i R|^2 \leq -\Delta|\nabla^i R|^2 + \sum_{j=1}^{i-1} C_{ij}|\nabla^j R||\nabla^{i-j}R||\nabla^i R| + C_{i0}\left(|R| + \frac{1}{\tau}\right)|\nabla^i R|^2$$

for the Riemann tensor. For $\mathrm{Ric} - \frac{1}{2\tau}g$, we have

$$\begin{aligned}\partial_t\left|\mathrm{Ric} - \frac{1}{2\tau}g\right|^2 &\leq -\Delta\left|\mathrm{Ric} - \frac{1}{2\tau}g\right|^2 + C\left(|R||\mathrm{Ric}| + \left|\frac{\partial_t\tau}{2\tau}\right|\right)\left|\mathrm{Ric} - \frac{1}{2\tau}g\right| \\ &\leq -\Delta\left|\mathrm{Ric} - \frac{1}{2\tau}g\right|^2 + C\left(|R||\mathrm{Ric}| + \frac{1}{2\tau}\left|\mathrm{Ric} - \frac{1}{2\tau}g\right|\right)\left|\mathrm{Ric} - \frac{1}{2\tau}g\right|,\end{aligned}$$

where we used Lemma 6.3.2 for the estimate $|\partial_t\tau| \leq C|\mathrm{Ric} - \frac{1}{2\tau}g|$. For higher derivatives, we have

$$\begin{aligned}\partial_t\left|\nabla^i\left(\mathrm{Ric} - \frac{1}{2\tau}g\right)\right|^2 \leq &-\Delta\left|\nabla^i\left(\mathrm{Ric} - \frac{1}{2\tau}g\right)\right|^2 \\ &+ \sum_{j=0}^{i}\tilde{C}_{ij}|\nabla^j R||\nabla^{i-j}\mathrm{Ric}|\left|\nabla^i\left(\mathrm{Ric} - \frac{1}{2\tau}g\right)\right|.\end{aligned}$$

The rest of the proof is exactly as in Lemma 5.4.10 and uses the maximum principle for scalars. □

Lemma 6.5.6. *Let $g(t)$, $t \in [0,T]$ be a solution of the Ricci flow (6.32) and suppose that*

$$\sup_{p\in M}|R_{g(t)}|_{g(t)} + \frac{1}{\tau_{g(t)}} \leq T^{-1} \qquad \forall t \in [0,T].$$

Then for each $k \geq 1$, there exists a constant $C(k)$ such that

$$\sup_{p\in M}|\nabla^k R_{g(t)}|_{g(t)} \leq C(k)\cdot T^{-1}t^{-k/2} \qquad \forall t \in (0,T].$$

Proof. By the evolution equation $\partial_t R = -\Delta R + R * R + \frac{2}{\tau}R$, we have the evolution inequality

$$\begin{aligned}\partial_t|\nabla^i R|^2 \leq &-\Delta|\nabla^i R|^2 - 2|\nabla^{i+1}R|^2 + \sum_{j=1}^{i-1} C_{ij}|\nabla^j R||\nabla^{i-j}R||\nabla^i R| \\ &+ C_{i0}\left(|R| + \frac{1}{\tau}\right)|\nabla^i R|^2.\end{aligned}$$

The proof follows from induction on i exactly as in Lemma 5.4.11. □

Remark 6.5.7. As in Remark 5.4.12 for the flow (5.5), we obtain uniform bounds of all derivatives of the curvature along the Ricci flow (6.32) on $[\delta, T]$ if the curvature and $\frac{1}{\tau}$ are bounded on $[0,T]$.

Theorem 6.5.8 (Dynamical stability modulo diffeomorphism). *Let (M, g_E) be a compact positive Einstein manifold with constant μ and let $k \geq 3$. Suppose that g_E is a local maximizer of the Yamabe functional and the smallest nonzero eigenvalue of the Laplacian is larger than 2μ. Then for every C^k-neighbourhood $\mathcal{U}$ of g_E, there exists a C^{k+2}-neighbourhood $\mathcal{V}$ such that the following holds:*

For any metric $g_0 \in \mathcal{V}$, there exists a 1-parameter family of diffeomorphisms φ_t and a positive function v such that for the Ricci flow (6.32) starting at g_0, the modified flow $\varphi_t^* g(t)$ stays in $\mathcal{U}$ for all time and converges to an Einstein metric g_∞ in $\mathcal{U}$ as $t \to \infty$. The convergence is of polynomial rate, i.e. there exist constants $C, \alpha > 0$ such that

$$\|\varphi_t^* g(t) - g_\infty\|_{C^k} \leq C(t+1)^{-\alpha}.$$

Proof. Without loss of generality, we may assume that $\mathcal{U} = \mathcal{B}_\epsilon^k$ and that $\epsilon > 0$ is so small that Theorems 6.5.1 and 6.5.4 hold on $\mathcal{U}$.

By Lemma 6.5.5, we can choose a small neighbourhood $\mathcal{V}$ such that the Ricci flow, starting at any metric $g \in \mathcal{V}$ stays in $\mathcal{B}_{\epsilon/4}^k$ up to time 1. Let $T \geq 1$ be the maximal time such that for any Ricci flow $g(t)$ starting in $\mathcal{V}$, there exists a family of diffeomorphisms φ_t such that the modified flow $\varphi_t^* g(t)$ stays in $\mathcal{U}$. By definition of T and by diffeomorphism invariance, we have uniform curvature bounds

$$\sup_{p \in M} |R_{g(t)}|_{g(t)} \leq C_1 \qquad \forall t \in [0, T),$$
$$|\tau_{g(t)}| \leq C_2 \qquad \forall t \in [0, T).$$

By Remark 6.5.7, we have

$$\sup_{p \in M} |\nabla^l R_{g(t)}|_{g(t)} \leq C(l) \qquad \forall t \in [1, T). \tag{6.33}$$

Because $f_{g(t)}$ satisfies the equation $\tau(2\Delta f + |\nabla f|^2 - \mathrm{scal}) - f + n + \nu_- = 0$, we also have

$$\sup_{p \in M} |\nabla^l f_{g(t)}|_{g(t)} \leq \tilde{C}(l) \qquad \forall t \in [1, T). \tag{6.34}$$

Note that all these estimates are diffeomorphism invariant.

We now construct a modified Ricci flow as follows: Let $\varphi_t \in \mathrm{Diff}(M)$, $t \geq 1$ be the family of diffeomorphisms generated by $X(t) = -\mathrm{grad}_{g(t)} f_{g(t)}$ and define

$$\tilde{g}(t) = \begin{cases} g(t), & t \in [0,1], \\ \varphi_t^* g(t), & t \geq 1. \end{cases} \tag{6.35}$$

The modified flow satisfies the usual Ricci flow equation for $t \in [0, 1]$ while for $t \geq 1$, we have

$$\begin{aligned} \frac{d}{dt}\tilde{g}(t) &= \varphi_t^*(\dot{g}(t)) + \varphi_t^*(\mathcal{L}_{X(t)} g(t)) \\ &= \varphi_t^*\left(-2\mathrm{Ric}_{g(t)} + \frac{1}{\tau_{g(t)}} g(t)\right) - 2\varphi_t^*(\nabla^2 f_{g(t)}) \\ &= -2\mathrm{Ric}_{\tilde{g}(t)} + \frac{1}{\tau_{\tilde{g}(t)}}\tilde{g}(t) + \nabla^2 f_{\tilde{g}(t)}. \end{aligned}$$

Let $T' \in [0,T]$ be the maximal time such that the modified Ricci flow, starting at any metric $g_0 \in \mathcal{V}$, stays in $\mathcal{U}$ up to time t. Then

$$\begin{aligned}\|\tilde{g}(T') - g_E\|_{C^k} &\leq \|\tilde{g}(1) - g_E\|_{C^k} + \int_1^{T'} \|\dot{\tilde{g}}(t)\|_{C^k}\, dt \\ &\leq \frac{\epsilon}{4} + \int_1^{T'} \|\dot{\tilde{g}}(t)\|_{C^k}\, dt.\end{aligned}$$

By interpolation (c.f. [Ham82, Corollary 12.7]), (6.33) and (6.34), we have

$$\|\dot{\tilde{g}}(t)\|_{C^k} \leq C_3 \|\dot{\tilde{g}}(t)\|_{L^2}^{1-\eta}$$

for η as small as we want. In particular, we can assume that $\theta := 1-\sigma(1+\eta) > 0$, where σ is the constant appearing in the Lojasiewicz-Simon inequality 6.5.4. By the first variation of ν_-,

$$\frac{d}{dt}\nu_-(\tilde{g}(t)) \geq C_4 \|\dot{\tilde{g}}(t)\|_{L^2}^{1+\eta} \|\dot{\tilde{g}}(t)\|_{L^2}^{1-\eta}.$$

By Theorem 6.5.1 and again Theorem 6.5.4,

$$\begin{aligned}-\frac{d}{dt}|\nu_-(\tilde{g}(t)) - \nu_-(g_E)|^{\theta} &= \theta|\nu_-(\tilde{g}(t)) - \nu_-(g_E)|^{\theta-1}\frac{d}{dt}\nu_-(\tilde{g}(t)) \\ &\geq C_5|\nu_-(\tilde{g}(t)) - \nu_-(g_E)|^{-\sigma(1+\eta)} \|\dot{\tilde{g}}(t)\|_{L^2}^{1+\eta} \|\dot{\tilde{g}}(t)\|_{L^2}^{1-\eta} \\ &\geq C_6 \|\dot{\tilde{g}}(t)\|_{C^k}.\end{aligned}$$

Hence by integration,

$$\int_1^{T'} \|\dot{\tilde{g}}(t)\|_{C^k}\, dt \leq \frac{1}{C_6}|\nu_-(\tilde{g}(1)) - \nu_-(g_E)|^{\theta} \leq \frac{1}{C_6}|\nu_-(\tilde{g}(0)) - \nu_-(g_E)|^{\theta} \leq \frac{\epsilon}{4},$$

provided that $\mathcal{V}$ is small enough. Thus, $T = \infty$ and $\tilde{g}(t)$ converges to some limit metric $g_\infty \in \mathcal{U}$ as $t \to \infty$. By the Lojasiewicz-Simon inequality, we have

$$\frac{d}{dt}|\nu_-(\tilde{g}(t)) - \nu_-(g_E)|^{1-2\sigma} \geq C_7,$$

which implies

$$|\nu_-(\tilde{g}(t)) - \nu_-(g_E)| \leq C_8(t+1)^{-\frac{1}{2\sigma-1}}.$$

Therefore, $\nu_-(g_\infty) = \nu_-(g_E)$, so g_∞ is an Einstein metric by Theorem 6.5.1. The convergence is of polynomial rate, since for $t_1 < t_2$,

$$\|\tilde{g}(t_1) - \tilde{g}(t_2)\|_{C^k} \leq C_9|\nu_-(\tilde{g}(t_1)) - \nu_-(g_E)|^{\theta} \leq C_{10}(t_1+1)^{-\frac{\theta}{2\sigma-1}}.$$

The assertion follows from $t_2 \to \infty$. □

Theorem 6.5.9 (Dynamical instability modulo diffeomorphism). *Let (M, g_E) be a positive Einstein manifold that is not a local maximizer of ν_-. Then there exists a nontrivial ancient Ricci flow $g(t)$, $t \in (-\infty, 0]$ and a 1-parameter family of diffeomorphisms φ_t, $t \in (-\infty, 0]$ such that $\varphi_t^* g(t) \to g_E$ as $t \to \infty$.*

Proof. Let $g_i \to g_E$ in C^k and suppose that $\nu_-(g_i) > \nu_-(g_E)$ for all i. Let $\tilde{g}_i(t)$ be the modified flow defined in (6.35), which starts at g_i. Then by Lemma 6.5.5, $\bar{g}_i = g_i(1)$ converges to g_E in C^{k-2} and by monotonicity, $\nu_-(\bar{g}_i) > \nu_-(g_E)$ as well. Let $\epsilon > 0$ be so small that Theorem 6.5.4 holds on $\mathcal{B}^{k-2}_{2\epsilon}$. Theorem 6.4.2 yields the differential inequality

$$\frac{d}{dt}(\nu_-(\tilde{g}_i(t)) - \nu_-(g_E))^{1-2\sigma} \geq -C_1,$$

from which we obtain

$$[(\nu_-(\tilde{g}_i(t)) - \nu_-(g_E))^{1-2\sigma} - C_1(s-t)]^{-\frac{1}{2\sigma-1}} \leq (\nu_-(\tilde{g}_i(s)) - \nu_-(g_E)),$$

as long as $\tilde{g}_i(t)$ stays in $\mathcal{B}^{k-2}_{2\epsilon}$. Thus, there exists a t_i such that

$$\|\tilde{g}_i(t_i) - g_E\|_{C^{k-2}} = \epsilon,$$

and $t_i \to \infty$. If $\{t_i\}$ was bounded, $\tilde{g}_i(t_i) \to g_E$ in C^{k-2}. By interpolation,

$$\left\|\dot{\tilde{g}}_i(t)\right\|_{C^{k-2}} \leq C_2 \left\|\dot{\tilde{g}}_i(t)\right\|_{L^2}^{1-\eta}$$

for $\eta > 0$ as small as we want. We may assume that $\theta = 1 - \sigma(1+\eta) > 0$. By Theorem 6.5.4 , we have the differential inequality

$$\frac{d}{dt}(\nu_-(\tilde{g}_i(t)) - \nu_-(g_E))^{\theta} \geq C_3 \left\|\dot{\tilde{g}}_i(t)\right\|_{L^2}^{1-\eta},$$

if $\nu_-(\tilde{g}_i(t)) > \nu_-(g_E)$. Thus,

$$\epsilon = \|\tilde{g}_i(t_i) - g_E\|_{C^{k-2}} \leq \|\bar{g}_i - g_E\|_{C^{k-2}} + C_4(\nu_-(\tilde{g}_i(t_i)) - \nu_-(g_E))^{\theta}. \quad (6.36)$$

Now put $\tilde{g}^s_i(t) := \tilde{g}_i(t+t_i)$, $t \in [T_i, 0]$, where $T_i = 1 - t_i \to -\infty$. We have

$$\begin{aligned} \|\tilde{g}^s_i(t) - g_E\|_{C^{k-2}} \leq \epsilon \qquad & \forall t \in [T_i, 0], \\ \tilde{g}^s_i(T_i) \to g_E & \text{ in } C^{k-2}. \end{aligned}$$

Because the embedding $C^{k-3}(M) \subset C^{k-2}(M)$ is compact, we can choose a subsequence of the $\tilde{g}^s_i$, converging in $C^{k-3}_{loc}(M \times (-\infty, 0])$ to an ancient flow $\tilde{g}(t)$, $t \in (-\infty, 0]$, which satisfies the differential equation

$$\dot{\tilde{g}}(t) = -2\left(\mathrm{Ric}_{\tilde{g}(t)} - \frac{1}{2\tau_{\tilde{g}(t)}}\tilde{g}(t) + \nabla^2 f_{\tilde{g}(t)}\right).$$

Let φ_t, $t \in (-\infty, 0]$ be the diffeomorphisms generated by $X(t) = \mathrm{grad}_{\tilde{g}(t)} f_{\tilde{g}(t)}$, where $\varphi_0 = \mathrm{id}$. Then $g(t) = \varphi_t^* \tilde{g}(t)$ is a solution of (6.32). From taking the limit $i \to \infty$ in (6.36), we have $\epsilon \leq C_4(\nu_-(g(0)) - \nu_-(g_E))^{\beta/2}$ which shows that the Ricci flow is nontrivial. For $T_i \leq t$, the Lojasiewicz-Simon inequality implies

$$\begin{aligned} \|\tilde{g}^s_i(T_i) - \tilde{g}^s_i(t)\|_{C^{k-3}} \leq & C_4(\nu_-(\tilde{g}_i(t+t_i)) - \nu_-(g_E))^{\theta} \\ \leq & C_4[-C_1 t + (\nu_-(\tilde{g}_i(t_i)) - \nu_-(g_E))^{1-2\sigma}]^{-\frac{\theta}{2\sigma-1}} \\ \leq & [-C_5 t + C_6]^{-\frac{\theta}{2\sigma-1}}. \end{aligned}$$

Thus,

$$\begin{aligned}\|g_E - \tilde{g}(t)\|_{C^{k-3}} \leq &\|g_E - \tilde{g}_i^s(T_i)\|_{C^{k-3}} + [-C_5 t + C_6]^{-\frac{\theta}{2\sigma-1}} \\ &+ \|\tilde{g}_i^s(t) - \tilde{g}(t)\|_{C^{k-3}}.\end{aligned}$$

It follows that $\|g_E - \tilde{g}(t)\|_{C^{k-3}} \to 0$ as $t \to -\infty$. Therefore, $(\varphi_t^{-1})^* g(t) \to g_E$ in C^{k-3} as $t \to -\infty$ which proves the theorem. □

Remark 6.5.10. We hope to generalize Theorems 6.5.8 and 6.5.9 to the case of shrinking gradient Ricci solitons, i.e. we want to characterize dynamical stability and instability of them in terms of the local behavior of ν_-.

6.6 Dynamical Instability of the Complex Projective Space

Theorem 6.5.1 is rather unsatisfactory, because we cannot completely characterize the maximality of the shrinker entropy in terms of the local behavior of the Yamabe functional and an eigenvalue assumption. In fact there are several examples of Einstein manifolds (including $(\mathbb{C}P^n, g_{st})$, see [CH13]) which are local maxima of the Yamabe functional but to which we cannot apply Theorem 6.5.1 because 2μ (where μ is the Einstein constant) is exactly the smallest nonzero eigenvalue of the Laplacian.

In this section, we prove an instability criterion for such Einstein metrics. The idea is simple but its realization needs a long calculation. It consists of explicitly computing a third variation of the shrinker entropy.

Proposition 6.6.1. *Let (M, g_E) be a positive Einstein manifold with constant μ and suppose we have a function $v \in C^\infty(M)$ such that $\Delta v = 2\mu \cdot v$. Then the third variation of ν_- in the direction of $v \cdot g_E$ is given by*

$$\left.\frac{d^3}{dt^3}\right|_{t=0} \nu_-(g_E + tv \cdot g_E) = (n-2) \fint_M v^3 \, dV.$$

Proof. Put $u = \frac{e^{-f}}{(4\pi\tau)^{n/2}}$. By the first variation, the negative of the $L^2(u\,dV)$-gradient of ν_- is given by $\nabla\nu_- = \tau(\mathrm{Ric} + \nabla^2 f) - \frac{g}{2}$, so

$$\left.\frac{d}{dt}\right|_{t=0} \nu_-(g_E + th) = -\int_M \langle \nabla\nu_-, h\rangle u \, dV.$$

Since (M, g_E) is a critical point of ν_-, we clearly have $\nabla\nu_- = 0$. Since v is a nonconstant eigenfunction, $\int_M v \, dV = 0$. Thus by Lemma 6.2.3, τ' vanishes. Recall from (6.6) that $\tau_{g_E} = \frac{1}{2\mu}$ and f_{g_E} is constant. Therefore, by the first variation of the Ricci tensor,

$$\begin{aligned}\nabla\nu_-' =&\tau'\mu g_E + \frac{1}{2\mu}(\mathrm{Ric}' + \nabla^2(f')) - \frac{g'}{2} \\ =&\frac{1}{2\mu}\left(\frac{1}{2}\Delta_L(v\cdot g_E) - \delta^*\delta(v\cdot g_E) - \frac{1}{2}\nabla^2\mathrm{tr}(v\cdot g_E) + \nabla^2(f')\right) - \frac{v\cdot g_E}{2} \\ =&\frac{1}{2\mu}\left(\frac{1}{2}\Delta v\cdot g_E + (1-\frac{n}{2})\nabla^2 v + \nabla^2(f')\right) - \frac{v\cdot g_E}{2} \\ =&\frac{1}{2\mu}\left(\left(1-\frac{n}{2}\right)\nabla^2 v + \nabla^2(f')\right).\end{aligned}$$

To compute f', we consider the Euler-Lagrange equation

$$\tau(2\Delta f + |\nabla f|^2 - \mathrm{scal}) - f + n + \nu_- = 0. \tag{6.37}$$

By differentiating once and using $\tau' = 0$ and $\nu_-' = 0$,

$$\frac{1}{2\mu}(2\Delta f' - \mathrm{scal}') - f' = 0,$$

and by the first variation of the scalar curvature,

$$\begin{aligned}\left(\frac{1}{\mu}\Delta - 1\right) f' = \frac{1}{2\mu}\mathrm{scal}' =&\frac{1}{2\mu}(\Delta(\mathrm{tr}(v\cdot g_E)) + \delta\delta(v\cdot g_E) - \langle \mathrm{Ric}, v\cdot g_E\rangle) \\ =&\frac{1}{2\mu}((n-1)\Delta v - n\mu v).\end{aligned}$$

By Obata's eigenvalue estimate, $\frac{1}{\mu}\Delta - 1$ is invertible. By using the eigenvalue equation, we therefore obtain

$$f' = \left(\frac{n}{2} - 1\right) v. \tag{6.38}$$

Thus,

$$\nabla\nu_-' = 0, \tag{6.39}$$

and therefore, the third variation equals

$$\left.\frac{d^3}{dt^3}\right|_{t=0} \nu_-(g_E + tv\cdot g_E) = -\int_M \langle \nabla\nu_-'', v\cdot g_E\rangle u\, dV.$$

Since $\tau_{g_E} = \frac{1}{2\mu}$ and $\tau' = 0$,

$$\nabla\nu_-'' = -\tau''\cdot g_E + \frac{1}{2\mu}(\mathrm{Ric} + \nabla^2 f)''.$$

The function u is constant since f is constant. Thus, the τ''-term drops out after integration. We are left with

$$\left.\frac{d^3}{dt^3}\right|_{t=0} \nu_-(g_E + tv\cdot g_E) = -\frac{1}{2\mu}\int_M \langle (\mathrm{Ric} + \nabla^2 f)'', v\cdot g_E\rangle u\, dV. \tag{6.40}$$

We first compute Ric''. Let $g_t = (1+tv)g_E$ and $v_t = \frac{v}{1+tv}$. Then $g_t' = v_t \cdot g_t$ and $\frac{d}{dt}|_{t=0} v_t = -v^2$. By the first variation of the Ricci tensor,

$$\begin{aligned} \frac{d}{dt}\mathrm{Ric}_{g_t} &= \frac{1}{2}\Delta_L(v_t \cdot g_t) - \delta^*\delta(v_t \cdot g_t) - \frac{1}{2}\nabla^2 \mathrm{tr}(v_t \cdot g_t) \\ &= \frac{1}{2}[(\Delta v_t) g_t - (n-2)\nabla^2 v_t], \end{aligned}$$

and the second variation at g_E is equal to

$$\begin{aligned} \frac{d^2}{dt^2}\bigg|_{t=0} \mathrm{Ric}_{g_E + tv\cdot g_E} &= \frac{d}{dt}\bigg|_{t=0} \frac{1}{2}[(\Delta v_t) g_t - (n-2)\nabla^2 v_t] \\ =&\frac{1}{2}[(\Delta' v + \Delta(v') + \Delta v \cdot v) g_E - (n-2)(\nabla^2)' v - (n-2)\nabla^2(v')] \\ =&\frac{1}{2}[(\langle v \cdot g_E, \nabla^2 v\rangle - \langle \delta(v\cdot g_E) + \frac{1}{2}\nabla \mathrm{tr}(v \cdot g_E), \nabla v\rangle) g_E \\ &+ (-\Delta v \cdot v + 2|\nabla v|^2) g_E - (n-2)\left(\frac{1}{2}|\nabla v|^2 g_E - \nabla v \otimes \nabla v\right) \\ &+ (n-2)(2\nabla^2 v \cdot v + 2\nabla v \otimes \nabla v)] \\ =&-\left(\frac{n}{2}-2\right)|\nabla v|^2 g_E - (\Delta v \cdot v) g_E + 3\left(\frac{n}{2}-1\right)\nabla v \otimes \nabla v + (n-2)\nabla^2 v \cdot v \\ =&-\left(\frac{n}{2}-2\right)|\nabla v|^2 g_E - 2\mu v^2 g_E + 3\left(\frac{n}{2}-1\right)\nabla v \otimes \nabla v + (n-2)\nabla^2 v \cdot v, \end{aligned}$$

where we used the first variational formulas of the Laplacian and the Hessian in Lemma A.3. Let us now compute the $(\nabla^2 f)''$-term. Since f_{g_E} is constant,

$$\begin{aligned} \frac{d^2}{dt^2}\bigg|_{t=0} \nabla^2 f_{g_E + tv\cdot g_E} &= \nabla^2(f'') + 2(\nabla^2)' f' \\ &= \nabla^2(f'') - \nabla v \otimes \nabla f' - \nabla f' \otimes \nabla v + \langle \nabla f', \nabla v\rangle g_E. \end{aligned}$$

We already know that $f' = (\frac{n}{2}-1)v$ by (6.38). To compute f'', we differentiate (6.37) twice. By (6.39), $\nu_-'' = 0$. Since also $\tau' = 0$ as remarked above, we obtain

$$\begin{aligned} 0 &= -\tau'' \mathrm{scal} + \tau(2\Delta f + |\nabla f|^2 - \mathrm{scal})'' - f'' \\ &= -\tau'' n\mu + \frac{1}{\mu}\Delta f'' + \frac{2}{\mu}\Delta' f' + \frac{1}{\mu}|\nabla(f')|^2 - \frac{1}{2\mu}\mathrm{scal}'' - f''. \end{aligned} \tag{6.41}$$

Because $\Delta v = 2\mu v$,

$$\begin{aligned} \Delta' f' =& \langle v \cdot g, \nabla^2 f'\rangle - \left\langle \delta(v \cdot g) + \frac{1}{2}\nabla \mathrm{tr}(v \cdot g), \nabla f'\right\rangle \\ \overset{6.38}{=}& \left(\frac{n}{2}-1\right)\left[-v\Delta v - \langle -\nabla v + \frac{n}{2}\nabla v, \nabla v\rangle\right] \\ =& \left(\frac{n}{2}-1\right)\left[-2\mu v^2 - \left(\frac{n}{2}-1\right)|\nabla v|^2\right]. \end{aligned} \tag{6.42}$$

Next, we compute scal''. As above, let $g_t = (1+tv)g_E$ and $v_t = \frac{v}{1+tv}$. Then by the first variation of the scalar curvature,

$$\begin{aligned} \frac{d}{dt}\mathrm{scal}_{g_t} =& \Delta \mathrm{tr} g_t' + \delta\delta(g_t') - \langle \mathrm{Ric}_{g_t}, g_t'\rangle \\ =& (n-1)\Delta v_t - \mathrm{scal}_{g_t} v_t. \end{aligned}$$

The second variation of the scalar curvature at g_E is equal to

$$
\begin{aligned}
\frac{d^2}{dt^2}\bigg|_{t=0} \mathrm{scal}_{g_E+tv\cdot g_E} =& \frac{d}{dt}\bigg|_{t=0}[(n-1)\Delta v_t - \mathrm{scal}_{g_t} v_t] \\
=&(n-1)[\Delta' v + \Delta(v')] - n\mu\cdot v' - \mathrm{scal}' v \\
=&(n-1)[\langle v\cdot g_E, \nabla^2 v\rangle - \langle \delta(v\cdot g_E) + \frac{1}{2}\nabla \mathrm{tr}(v\cdot g_E), \nabla v\rangle - \Delta(v^2)] \\
&+ n\mu\cdot v^2 - [\Delta \mathrm{tr}(v\cdot g_E) + \delta\delta(v\cdot g_E) - \langle \mathrm{Ric}, v\cdot g_E\rangle]v \\
=& -(n-1)\left(\frac{n}{2}-3\right)|\nabla v|^2 + 2\mu(4-3n)\cdot v^2.
\end{aligned}
$$

By (6.38), $|\nabla(f')|^2 = (\frac{n}{2}-1)^2|\nabla v|^2$. Thus, we can rewrite (6.41) as

$$
\begin{aligned}
\left(\frac{1}{\mu}\Delta - 1\right) f'' =& \tau'' n\mu - \frac{1}{\mu}(2\Delta' f' + |\nabla(f')|^2 - \frac{1}{2}\mathrm{scal}'') \\
\overset{(6.42)}{=}& \tau'' n\mu - \frac{1}{\mu}[-2(n-2)\mu v^2 - 2\left(\frac{n}{2}-1\right)^2|\nabla v|^2 + \left(\frac{n}{2}-1\right)^2|\nabla v|^2 \\
&+ \frac{n-1}{2}\left(\frac{n}{2}-3\right)|\nabla v|^2 + (3n-4)\mu v^2] \\
=& \tau'' n\mu - \frac{1}{\mu}\left[n\mu v^2 + \left(-\frac{3}{4}n + \frac{1}{2}\right)|\nabla v|^2\right] =: (A).
\end{aligned}
$$

Since $\frac{1}{\mu}\Delta - 1$ is invertible, we can rewrite the above as

$$
f'' = (\frac{1}{\mu}\Delta - 1)^{-1}(A).
$$

By integrating,

$$
\begin{aligned}
-\frac{1}{2\mu}\int_M \langle(\nabla^2 f)'', v\cdot g_E\rangle u\, dV =& -\frac{1}{2\mu}\fint_M \langle(\nabla^2 f)'', v\cdot g_E\rangle\, dV \\
=& -\frac{1}{2\mu}\fint_M \langle \nabla^2(f'') - \nabla v\otimes\nabla f' - \nabla f'\otimes\nabla v + \langle\nabla f', \nabla v\rangle g_E, v\cdot g_E\rangle\, dV \\
\overset{6.38}{=}& -\frac{1}{2\mu}\fint_M \left\langle \nabla^2(f'') - (n-2)\nabla v\otimes\nabla v + \left(\frac{n}{2}-1\right)|\nabla v|^2 g_E, v\cdot g_E\right\rangle\, dV \\
=& -\frac{1}{2\mu}\fint_M \left[-\Delta(f'')v + \frac{1}{2}(n-2)^2|\nabla v|^2 v\right]\, dV \\
=& -\frac{1}{2\mu}\fint_M \left[-(A)\left(\frac{1}{\mu}\Delta - 1\right)^{-1}\Delta v + \frac{1}{2}(n-2)^2|\nabla v|^2 v\right]\, dV \\
=& -\frac{1}{2\mu}\fint_M \left[-2\mu(A)v + \frac{1}{2}(n-2)^2|\nabla v|^2 v\right]\, dV.
\end{aligned}
$$

Now we insert the definition of (A). Since the term containing τ'' drops out after integration, we are left with

$$
-\frac{1}{2\mu}\int_M \langle(\nabla^2 f)'', v\cdot g_E\rangle u\, dV = -\frac{1}{2\mu}\fint_M \left[(2n\mu v^3 + \frac{1}{2}(n^2 - 7n + 6)|\nabla v|^2 v\right]\, dV.
$$

By the second variation of the Ricci tensor computed above,

$$\begin{aligned}
-\frac{1}{2\mu}\int_M \langle \mathrm{Ric}'', v\cdot g_E\rangle u\, dV &= -\frac{1}{2\mu}\fint_M \langle \mathrm{Ric}'', v\cdot g_E\rangle\, dV \\
&= -\frac{1}{2\mu}\fint_M [-n\left(\frac{n}{2}-2\right)|\nabla v|^2 v \\
&\quad - 2\mu n v^3 + 3\left(\frac{n}{2}-1\right)|\nabla v|^2 v - (n-2)\Delta v\cdot v^2]\, dV \\
&= -\frac{1}{2\mu}\fint_M \left[\left(-\frac{n^2}{2}+\frac{7n}{2}-3\right)|\nabla v|^2 v - 4(n-1)\mu v^3\right]\, dV.
\end{aligned}$$

Adding up these two terms, we obtain

$$\left.\frac{d^3}{dt^3}\right|_{t=0} \nu_-(g+tv\cdot g) \overset{(6.40)}{=} -\frac{1}{2\mu}\fint_M (4-2n)\mu v^3\, dV.$$

and therefore, we finally have

$$\left.\frac{d^3}{dt^3}\right|_{t=0} \nu_-(g+tv\cdot g) = (n-2)\fint_M v^3\, dV,$$

which finishes the proof. □

Corollary 6.6.2. *Let (M, g_E) be a positive Einstein manifold with constant μ. Suppose there exists a function $v \in C^\infty(M)$ such that $\Delta v = 2\mu v$ and $\int_M v^3\, dV \neq 0$. Then g_E is not a local maximum of ν_-.*

Proof. Let $\varphi(t) = \nu_-(g_E + tv\cdot g_E)$. By the proof of the proposition above, $\varphi'(0) = 0$, $\varphi''(0) = 0$ and $\varphi'''(0) \neq 0$. Depending on the sign of the third variation, $\varphi(t) > \varphi(0)$ either for $t \in (-\epsilon, 0)$ or $t \in (0, \epsilon)$. This proves the assertion. □

Because the eigenfunctions on $\mathbb{C}P^n$ can be constructed explicitly, we are able to find an eigenfunction satisfying the above condition. Thus we obtain

Theorem 6.6.3. *The manifold $(\mathbb{C}P^n, g_{st})$, $n > 1$ is dynamically unstable modulo diffeomorphism.*

Proof. Let μ be the Einstein constant. We prove the existence of a function $v \in C^\infty(\mathbb{C}P^n)$ satisfying $\Delta v = 2\mu v$ and $\int_{\mathbb{C}P^n} v^3\, dV \neq 0$. First, we rewiev the construction of eigenfunctions on $\mathbb{C}P^n$ as explained in [BGM71, Section III C]. Consider $\mathbb{C}^{n+1} = \mathbb{R}^{2n+2}$ with coordinates $(x_1, \dots, x_{n+1}, y_1, \dots, y_{n+1})$ and let $z_j = x_j + iy_j$, $\bar{z}_j = x_j - iy_j$ be the complex coordinates. Defining $\partial_{z_j} = \frac{1}{2}(\partial_{x_j} - i\partial_{y_j})$ and $\partial_{\bar{z}_j} = \frac{1}{2}(\partial_{x_j} - i\partial_{y_j})$, we can rewrite the Laplace operator on $\mathbb{C}^{n+1}$ as

$$\Delta = -4\sum_{j=1}^{n+1} \partial_{z_j} \circ \partial_{\bar{z}_j}.$$

Let $P_{k,k}$ be the space of complex polynomials on $\mathbb{C}^{n+1}$ which are homogeneous of degree k in z and $\bar{z}$ and let $H_{k,k}$ the subspace of harmonic polynomials in $P_{k,k}$. We have

$$P_{k,k} = H_{k,k} \oplus r^2 P_{k-1,k-1}.$$

Elements in $P_{k,k}$ are S^1-invariant and thus, they descend to functions on the quotient $\mathbb{C}P^n = S^{2n+1}/S^1$. The eigenfunctions to the k-th eigenvalue of the Laplacian on $\mathbb{C}P^n$ (where 0 is meant to be the 0-th eigenvalue) are precisely the restrictions of functions in $H_{k,k}$. Since 2μ is the first nonzero eigenvalue, its eigenfunctions are restrictions of functions in $H_{1,1}$.

Let $h_1(z,\bar{z}) = z_1\bar{z}_2 + z_2\bar{z}_1$, $h_2(z,\bar{z}) = z_2\bar{z}_3 + z_3\bar{z}_2$, $h_3(z,\bar{z}) = z_3\bar{z}_1 + z_1\bar{z}_3$ and let v be the eigenfunction which is the restriction of $h = h_1 + h_2 + h_3 \in H_{1,1}$. Note that h is real-valued and so is v. Then v^3 is the restriction of

$$h^3 \in P_{3,3} = H_{3,3} \oplus r^2 H_{2,2} \oplus r^4 H_{1,1} \oplus r^6 H_{0,0}. \tag{6.43}$$

We show that $\int_{S^{2n+1}} h^3 \, dV \neq 0$. At first,

$$h^3 = \sum_{j=1}^{3} h_j^3 + 3\sum_{j\neq l} h_j \cdot h_l^2 + 6h_1 \cdot h_2 \cdot h_3.$$

Note that $\int_{S^{2n+1}} h_1^3 \, dV = 0$ because h_1 is antisymmetric with respect to the isometry $(z_1, \bar{z}_1) \mapsto (-z_1, -\bar{z}_1)$. For the same reason, $\int_{S^{2n+1}} h_1 \cdot h_2^2 \, dV = 0$. Similarly, we show that all other terms of this form vanish after integration so it remains to deal with the last term of above. Note that

$$h_1 \cdot h_2 \cdot h_3(z,\bar{z}) = 2|z_1|^2|z_2|^2|z_3|^2 + \sum_{\sigma \in S_3} |z_{\sigma(1)}|^2 z_{\sigma(2)}^2 \bar{z}_{\sigma(3)}^2.$$

Consider $|z_1|^2 z_2^2 \bar{z}_3^2$. This polynomial is antisymmetric with respect to the isometry $(z_2, \bar{z}_2) \mapsto (i \cdot z_2, i \cdot \bar{z}_2)$ and therefore,

$$\int_{S^{2n+1}} |z_1|^2 z_2^2 \bar{z}_3^2 \, dV = 0.$$

Similarly, we deal with the other summands. In summary, we have

$$\int_{S^{2n+1}} h^3 \, dV = 6\int_{S^{2n+1}} h_1 \cdot h_2 \cdot h_3 \, dV = 12\int_{S^{2n+1}} |z_1|^2|z_2|^2|z_3|^2 \, dV > 0,$$

since the integrand on the right hand side is nonnegative and not identically zero. We decompose $h^3 = \sum_{j=0}^{3} h_j$, where $h_j \in r^{6-2j} H_{j,j}$. Since the restrictions of the h_j to S^{2n+1} are eigenfunctions to the $2j$-th eigenvalue of the Laplacian on S^{2n+1} (see [BGM71, Section III C]), we have that $h_0 \neq 0$ because the integral is nonvanishing. This decomposition induces a decomposition of $v^3 = \sum_{i=0}^{3} v_i$ where v_i is an eigenfunction of the i-th eigenvalue of $\Delta_{\mathbb{C}P^n}$ and $v_0 \neq 0$. Therefore, $\int_{\mathbb{C}P^n} v^3 \, dV \neq 0$.

By Corollary 6.6.2, $(\mathbb{C}P^n, g_{st})$ is not a local maximum of ν_- and thus, it is dynamically unstable modulo diffeomorphism by Theorem 6.5.9.

□

Remark 6.6.4. In contrast to the above, $(\mathbb{C}P^n, g_{st})$ is dynamically stable with respect to the Kähler-Ricci flow, see [SW13].

Remark 6.6.5. It is conjectured (c.f. [Cao10]) that the only linearly stable simply-connected4-dimensional positive Einstein manifolds are (S^n, g_{st}) and $(\mathbb{C}P^n, g_{st})$. If this conjecture holds, the above theorem implies that the round sphere is the only dynamically stable Einstein manifold in this class.

Remark 6.6.6. There are some other neutrally linearly stable Einstein metrics where $2\mu \in \operatorname{spec}(\Delta)$, see [CH13]. It seems likely that there we can find eigenfunctions with eigenvalue 2μ such that $\int_M v^3 \, dV \neq 0$.

Appendix A

Calculus of Variation

Here, we prove the variational formulas we used throughout the thesis.

Lemma A.1. *Let $\omega, \xi \in \Omega^1(M)$ and $T, S \in \Gamma(S^2M)$. Then the first variation of the induced scalar products and the trace are given by*

$$\frac{d}{dt}\bigg|_{t=0} \langle \omega, \xi \rangle_{g+th} = -h(\omega^\sharp, \xi^\sharp),$$
$$\frac{d}{dt}\bigg|_{t=0} \langle T, S \rangle_{g+th} = -2\langle T, h \circ S \rangle_g,$$
$$\frac{d}{dt}\bigg|_{t=0} \mathrm{tr}_{g+th} T = -\langle T, h \rangle_g.$$

Furthermore, the first variation of the volume element is given by

$$\frac{d}{dt}\bigg|_{t=0} dV_{g+th} = \frac{1}{2}\mathrm{tr}_g h \cdot \, dV_g.$$

Proof. We use local coordinates. The first formula follows from

$$\frac{d}{dt}\bigg|_{t=0} (g+th)^{ij}\omega_i\xi_j = -h^{ij}\omega_i\xi_j = -h_{kl}g^{ki}\omega_i g^{lj}\xi_j.$$

The second formula follows from

$$\begin{aligned}\frac{d}{dt}\bigg|_{t=0} (g+th)^{ij}(g+th)^{kl}T_{ik}S_{jl} &= -h^{ij}g^{kl}T_{ik}S_{jl} - g^{ij}h^{kl}T_{ik}S_{jl} \\ &= -2g^{ij}g^{km}T_{ik}S_{jl}g^{ln}h_{nm} \\ &= -2g^{ij}g^{km}T_{ik}(h \circ S)_{km}.\end{aligned}$$

The variation of the trace follows from

$$\frac{d}{dt}\bigg|_{t=0} (g+th)^{ij}T_{ij} = -h^{ij}T_{ij} = -g^{ki}g^{lj}h_{kl}T_{ij}.$$

Finally, we compute the first variation of the volume element and we obtain

$$\begin{aligned}\frac{d}{dt}\bigg|_{t=0} dV =& \frac{d}{dt}\bigg|_{t=0} [\det((g+th)_{ij})]^{1/2}\mathrm{dx} \\ =& \frac{1}{2}\det(g_{ij})^{-1/2}\frac{d}{dt}\bigg|_{t=0}[\det((g+th)_{ij})]\mathrm{dx} \\ =& \frac{1}{2}\mathrm{tr}h\det(g_{ij})^{1/2}\mathrm{dx} = \frac{1}{2}\mathrm{tr}h\cdot\, dV. \qquad \square\end{aligned}$$

Lemma A.2. *Let (M,g) be Riemannian manifold and denote the first variation of the Levi-Civita connection in the direction of h by G. Then G is a $(1,2)$ tensor field, given by*

$$g(G(X,Y),Z) = \frac{1}{2}(\nabla_X h(Y,Z) + \nabla_Y h(X,Z) - \nabla_Z h(X,Y)).$$

The first variation of the Riemann curvature tensor (as a $(1,3)$ and as a $(0,4)$-tensor), the Ricci tensor and the scalar curvature are given by

$$\begin{aligned}\frac{d}{dt}\bigg|_{t=0} {}^{g+th}R_{X,Y}Z =& (\nabla_X G)(Y,Z) - (\nabla_Y G)(X,Z), \\ \frac{d}{dt}\bigg|_{t=0} R_{g+th}(X,Y,Z,W) =& \frac{1}{2}(\nabla^2_{X,Z}h(Y,W) + \nabla^2_{Y,W}h(X,Z) - \nabla^2_{Y,Z}h(X,W) \\ & - \nabla^2_{X,W}h(Y,Z) + h(R_{X,Y}Z,W) - h(Z,R_{X,Y}W)), \\ \frac{d}{dt}\bigg|_{t=0} \mathrm{Ric}_{g+th}(X,Y) =& \frac{1}{2}\Delta_L h(X,Y) - \delta^*(\delta h)(X,Y) - \frac{1}{2}\nabla^2_{X,Y}\mathrm{tr}h, \\ \frac{d}{dt}\bigg|_{t=0} \mathrm{scal}_{g+th} =& \Delta_g(\mathrm{tr}_g h) + \delta_g(\delta_g h) - \langle \mathrm{Ric}_g, h\rangle_g.\end{aligned}$$

Proof. The difference between two connections is a $(1,2)$-tensor field, so G is. We do the computations at some point p and use normal coordinates with respect to g centered at p. First, we have

$$\begin{aligned}G^k_{ij} = \frac{d}{dt}\bigg|_{t=0}\Gamma^k_{ij} &= \frac{1}{2}g^{kl}(\partial_i h_{jl} + \partial_j h_{il} - \partial_k h_{ij}) \\ &= \frac{1}{2}g^{kl}(\nabla_i h_{jl} + \nabla_j h_{il} - \nabla_k h_{ij}).\end{aligned}$$

For the $(1,3)$ curvature tensor,

$$\begin{aligned}\frac{d}{dt}\bigg|_{t=0} R_{ijk}{}^{l} &= \frac{d}{dt}\bigg|_{t=0}(\partial_i\Gamma^l_{jk} - \partial_j\Gamma^l_{ik} + \Gamma^m_{jk}\Gamma^l_{im} - \Gamma^m_{ik}\Gamma^l_{jm}) \\ &= \partial_i G^l_{jk} - \partial_j G^l_{ik} \\ &= \nabla_i G^l_{jk} - \nabla_j G^l_{ik}.\end{aligned}$$

For the $(0,4)$ curvature tensor,

$$\begin{aligned}\frac{d}{dt}\bigg|_{t=0} R_{ijkl} = \frac{d}{dt}\bigg|_{t=0}(g_{lm}R_{ijk}{}^{m}) =& h_{lm}R_{ijk}{}^{m} + g_{lm}(\nabla_i G^m_{jk} - \nabla_j G^m_{ik}) \\ =& h_{lm}R_{ijk}{}^{m} + \frac{1}{2}(\nabla^2_{ij}h_{kl} + \nabla^2_{ik}h_{jl} - \nabla^2_{il}h_{jk}) \\ & - \frac{1}{2}(\nabla^2_{ji}h_{kl} + \nabla^2_{jk}h_{il} - \nabla^2_{jl}h_{ik}).\end{aligned}$$

By the Ricci identity,

$$\frac{1}{2}(\nabla^2_{ij}h_{kl} - \nabla^2_{ji}h_{kl}) = -\frac{1}{2}(R_{ijk}{}^m h_{ml} + R_{ijl}{}^m h_{km}),$$

which yields

$$\left.\frac{d}{dt}\right|_{t=0} R_{ijkl} = \frac{1}{2}(\nabla^2_{ik}h_{jl} - \nabla^2_{il}h_{jk} - \nabla^2_{jk}h_{il} + \nabla^2_{jl}h_{ik} + h_{lm}R_{ijk}{}^m - R_{ijl}{}^m h_{km}).$$

The first variation of the Ricci tensor is

$$\begin{aligned}\left.\frac{d}{dt}\right|_{t=0} \mathrm{Ric}_{jk} = \left.\frac{d}{dt}\right|_{t=0} R_{ijk}{}^i =& \nabla_i G^i_{jk} - \nabla_j G^i_{ik} \\ =& \frac{1}{2}g^{im}(\nabla^2_{ij}h_{km} + \nabla^2_{ik}h_{jm} - \nabla^2_{im}h_{jk}) \\ &- \frac{1}{2}g^{im}(\nabla^2_{ji}h_{km} + \nabla^2_{jk}h_{im} - \nabla^2_{jm}h_{ik}).\end{aligned}$$

Again by the Ricci identity,

$$\frac{1}{2}g^{im}(\nabla^2_{ij}h_{km} - \nabla^2_{ji}h_{km}) = -\frac{1}{2}g^{im}R_{ijk}{}^n h_{nm} + \frac{1}{2}\mathrm{Ric}^n_j h_{kn},$$

and

$$\begin{aligned}\frac{1}{2}g^{im}\nabla^2_{ik}h_{jm} = \frac{1}{2}g^{im}\nabla^2_{ik}h_{mj} &= \frac{1}{2}g^{im}(\nabla^2_{ik}h_{mj} - \nabla^2_{ki}h_{mj} + \nabla^2_{ki}h_{mj}) \\ &= \frac{1}{2}g^{im}(R_{kim}{}^n h_{nj} + R_{kij}{}^n h_{mn}) + \frac{1}{2}g^{im}\nabla^2_{ki}h_{mj} \\ &= \frac{1}{2}\mathrm{Ric}^m_k h_{mj} + \frac{1}{2}g^{im}R_{kij}{}^n h_{mn} - \frac{1}{2}\nabla_k(\delta h)_j.\end{aligned}$$

By rearranging the terms from above,

$$\begin{aligned}\left.\frac{d}{dt}\right|_{t=0} \mathrm{Ric}_{jk} =& \frac{1}{2}(-g^{im}\nabla^2_{im}h_{jk} + \mathrm{Ric}^m_k h_{mj} + \mathrm{Ric}^m_j h_{km} - 2g^{im}R^n_{ijk}h_{nm}) \\ &- \frac{1}{2}(\nabla_k(\delta h)_j + \nabla_j(\delta h)_k) - \frac{1}{2}\nabla_{jk}\mathrm{tr}h \\ =& \frac{1}{2}\Delta_L h_{jk} - \delta^*(\delta h)_{jk} - \frac{1}{2}\nabla^2_{jk}\mathrm{tr}h.\end{aligned}$$

The first variation of the scalar curvature is

$$\begin{aligned}\left.\frac{d}{dt}\right|_{t=0} \mathrm{scal} = \left.\frac{d}{dt}\right|_{t=0} (g^{ij}\mathrm{Ric}_{ij}) =& -h^{ij}\mathrm{Ric}_{ij} + g^{ij}\left.\frac{d}{dt}\right|_{t=0} \mathrm{Ric}_{ij} \\ =& -h^{ij}\mathrm{Ric}_{ij} + \Delta\mathrm{tr}h + \delta(\delta h).\end{aligned}$$ □

Lemma A.3. *The first variation of the Hessian and the Laplacian are given by*

$$\begin{aligned}\left.\frac{d}{dt}\right|_{t=0} {}^{g+th}\nabla^2_{X,Y}f =& -\frac{1}{2}[\nabla_X h(Y, \mathrm{grad}f) + \nabla_Y h(X, \mathrm{grad}f) - \nabla_{\mathrm{grad}f}h(X,Y)], \\ \left.\frac{d}{dt}\right|_{t=0} \Delta_{g+th}f =& \langle h, \nabla^2 f\rangle - \left\langle \delta h + \frac{1}{2}\nabla\mathrm{tr}h, \nabla f\right\rangle.\end{aligned}$$

The first variation of the symmetrised covariant differential and the divergence of a 1-form ω are given by

$$\left.\frac{d}{dt}\right|_{t=0} \delta^*_{g+th}\omega(X,Y) = -\frac{1}{2}[\nabla_X h(Y,\omega^\sharp) + \nabla_Y h(X,\omega^\sharp) - \nabla_{\omega^\sharp} h(X,Y)],$$

$$\left.\frac{d}{dt}\right|_{t=0} \delta_{g+th}\omega = \langle h, \nabla\omega\rangle - \delta h(\omega^\sharp) - \frac{1}{2}\langle \nabla \mathrm{tr} h, \omega\rangle.$$

Proof. We first prove the last two formulas. We again use local coordinates. Let ω be a 1-form. Then the first variation of its symmetrised covariant differential is given by

$$\begin{aligned}\left.\frac{d}{dt}\right|_{t=0} (\delta^*\omega)_{ij} &= \left.\frac{d}{dt}\right|_{t=0} \frac{1}{2}(\partial_i\omega_j + \partial_j\omega_i - (\Gamma^k_{ij} + \Gamma^k_{ji})\omega_k)\\ &= -\frac{1}{2}g^{kl}(\nabla_i h_{jl} + \nabla_j h_{il} - \nabla_l h_{ij})\omega_k\end{aligned}$$

by the first variation of the Levi-Civita connection. The first variation of the divergence is given by

$$\begin{aligned}\left.\frac{d}{dt}\right|_{t=0} (\delta\omega) &= -\left.\frac{d}{dt}\right|_{t=0} (g^{ij}(\delta^*\omega)_{ij})\\ &= h^{ij}(\delta^*\omega)_{ij} + \frac{1}{2}g^{ij}g^{kl}(\nabla_i h_{jl} + \nabla_j h_{il} - \nabla_l h_{ij})\omega_k\\ &= h^{ij}(\nabla\omega)_{ij} - g^{kl}(\delta h_l + \nabla_l \mathrm{tr} h)\omega_k.\end{aligned}$$

Since $\nabla^2 f = \delta^*(\nabla f)$ and $\Delta f = \delta(\nabla f)$, the first two formulas follow from the others by putting $\omega = \nabla f$. □

Lemma A.4. *Let h be a $(0,2)$-tensor field. Then we have*

$$\left.\frac{d}{dt}\right|_{t=0} \nabla_{g+tk} h = k * \nabla h + \nabla k * h,$$

$$\left.\frac{d}{dt}\right|_{t=0} \delta_{g+tk} h = k * \nabla h + \nabla k * h,$$

$$\left.\frac{d}{dt}\right|_{t=0} (\Delta_L)_{g+tk} h = k * \nabla^2 h + \nabla^2 k * h + \nabla k * \nabla h + R * k * h,$$

$$\left.\frac{d}{dt}\right|_{t=0} (\Delta_E)_{g+tk} h = k * \nabla^2 h + \nabla^2 k * h + \nabla k * \nabla h + R * k * h.$$

Here, $*$ is Hamilton's notation for a combination of tensor products with contractions.

Proof. The variation of the covariant differential of a $(0,2)$-tensor field h in the direction of k is given by

$$\begin{aligned}\left.\frac{d}{dt}\right|_{t=0} \nabla_i h_{jk} &= \left.\frac{d}{dt}\right|_{t=0} (\partial_i h_{jk} - \Gamma^l_{ij} h_{lk} - \Gamma^l_{ik} h_{jl})\\ &= -\frac{1}{2}g^{lm}(\nabla_i k_{jm} + \nabla_j k_{im} - \nabla_m k_{ij})h_{lk}\\ &\quad -\frac{1}{2}g^{lm}(\nabla_i k_{km} + \nabla_k k_{im} - \nabla_m k_{ik})h_{jl},\end{aligned}$$

which yields

$$\begin{aligned}
\frac{d}{dt}\Big|_{t=0}\delta h_k =& -\frac{d}{dt}\Big|_{t=0}(g^{ij}\nabla_i h_{jk})\\
=& h^{ij}\nabla_i h_{jk} + g^{ij}\frac{d}{dt}|_{t=0}\nabla_i h_{jk}\\
=& k^{ij}\nabla_i h_{jk} - \frac{1}{2}g^{ij}g^{lm}(\nabla_i k_{jm} + \nabla_j k_{im} - \nabla_m k_{ij})h_{lk}\\
& - \frac{1}{2}g^{ij}g^{lm}(\nabla_i k_{km} + \nabla_k k_{im} - \nabla_m k_{ik})h_{jl}.
\end{aligned}$$

To compute the last two formulas, we first compute the first variation of the Hessian on $(0, 2)$-tensors. Schematically, the local expression is of the form

$$\nabla^2_{ij}h_{kl} = \partial_i(\partial_j h_{kl} + (\Gamma * h)_{jkl}) + (\Gamma * \nabla h)_{ijkl}.$$

We now use normal coordinates with respect to g centered at some fixed point p. Then

$$\begin{aligned}
\frac{d}{dt}\Big|_{t=0}(\nabla^2_{ij}h_{kl}) &= \partial_i\left(\left(\frac{d}{dt}\Big|_{t=0}\Gamma\right) * h\right)_{jkl} + \left(\left(\frac{d}{dt}|_{t=0}\Gamma\right) * \nabla h\right)_{ijkl}\\
&= \nabla_i((\nabla k * h)_{jkl}) + (\nabla k * \nabla h)_{ijkl}\\
&= (\nabla^2 k * h)_{ijkl} + (\nabla k * \nabla h)_{ijkl}.
\end{aligned}$$

For the connection Laplacian, we have

$$\begin{aligned}
\frac{d}{dt}\Big|_{t=0}(\nabla^*\nabla h)_{kl} &= -\frac{d}{dt}\Big|_{t=0}(g^{ij}\nabla^2_{ij}h_{kl})\\
&= k^{ij}\nabla^2_{ij}h_{kl} - g^{ij}\frac{d}{dt}|_{t=0}(\nabla^2_{ij}h_{kl})\\
&= (k * \nabla^2 h)_{kl} + (\nabla^2 k * h)_{kl} + (\nabla k * \nabla h)_{kl}.
\end{aligned}$$

By Lemma A.2, the first variational formulas for the Riemann curvature tensor and the Ricci tensor are of the form

$$\begin{aligned}
\frac{d}{dt}\Big|_{t=0} R_{ijkl} &= (\nabla^2 * h)_{ijkl} + (R * h)_{ijkl},\\
\frac{d}{dt}\Big|_{t=0} \mathrm{Ric}_{ij} &= \frac{1}{2}\Delta_L h_{ij} - \delta^*(\delta h)_{ij} - \frac{1}{2}\nabla^2_{ij}(\mathrm{tr}h),\\
&= (\nabla^2 * h)_{ij} + (R * h)_{ij}.
\end{aligned}$$

Therefore, the variation of the Lichnerowicz Laplacian in the direction of k is given by

$$\begin{aligned}
\frac{d}{dt}\Big|_{t=0}(\Delta_L h)_{ij} &= \frac{d}{dt}\Big|_{t=0}(\nabla^*\nabla h - \mathrm{Ric}\circ h - h\circ \mathrm{Ric} - 2\mathring{R}h)_{ij}\\
&= (k * \nabla^2 h)_{ij} + (\nabla^2 k * h)_{ij} + (\nabla k * \nabla h)_{ij} + (R * k * h)_{ij}.
\end{aligned}$$

Similarly,

$$\begin{aligned}
\frac{d}{dt}\Big|_{t=0}(\Delta_E h)_{ij} =& \frac{d}{dt}\Big|_{t=0}(\nabla^*\nabla h - 2\mathring{R}h)_{ij}\\
=& (k * \nabla^2 h)_{ij} + (\nabla^2 k * h)_{ij} + (\nabla k * \nabla h)_{ij} + (R * k * h)_{ij}. \square
\end{aligned}$$

Lemma A.5. *The second variations of the Hessian, the Laplacian, the Ricci tensor and the scalar curvature have the schematic expressions*

$$\begin{aligned}
\frac{d}{ds}\frac{d}{dt}\Big|_{s,t=0}\nabla^2_{g+sk+th}f =&k*\nabla h*\nabla f+\nabla k*h*\nabla f,\\
\frac{d}{ds}\frac{d}{dt}\Big|_{s,t=0}\Delta_{g+sk+th}f =&k*\nabla h*\nabla f+\nabla k*h*\nabla f,\\
\frac{d}{ds}\frac{d}{dt}\Big|_{s,t=0}\mathrm{Ric}_{g+sk+th} =&k*\nabla^2 h+\nabla^2 k*h+\nabla k*\nabla h+R*k*h,\\
\frac{d}{ds}\frac{d}{dt}\Big|_{s,t=0}\mathrm{scal}_{g+sk+th} =&k*\nabla^2 h+\nabla^2 k*h+\nabla k*\nabla h+R*k*h.
\end{aligned}$$

Proof. We first compute the second variation of the Hessian in the direction of h and k. Using Lemma A.3 and Lemma A.4, we obtain

$$\begin{aligned}
\frac{d}{ds}\frac{d}{dt}\Big|_{s,t=0}\nabla^2_{ij}f &= \frac{d}{ds}\Big|_{s=0}\left(-\frac{1}{2}g^{kl}(\nabla_i h_{jl}+\nabla_j h_{il}-\nabla_l h_{ij})\partial_k f\right)\\
&=\frac{1}{2}k^{kl}(\nabla_i h_{jl}+\nabla_j h_{il}-\nabla_l h_{ij})\partial_k f)+\nabla k*h*\nabla f\\
&=k*\nabla h*\nabla f+\nabla k*h*\nabla f.
\end{aligned}$$

Therefore, using Lemma A.3 again, the second variation of the Laplacian is given by

$$\begin{aligned}
\frac{d}{ds}\frac{d}{dt}\Big|_{s,t=0}\Delta f &= -\frac{d}{ds}\frac{d}{dt}\Big|_{s,t=0}(g^{ij}\nabla^2_{ij}f)\\
&=h^{ij}\frac{d}{ds}\Big|_{s=0}\nabla^2_{ij}f+k^{ij}\frac{d}{dt}\Big|_{t=0}\nabla^2_{ij}f-g^{ij}\frac{d}{ds}\frac{d}{dt}\Big|_{s,t=0}\nabla^2_{ij}f\\
&=k*\nabla h*\nabla f+\nabla k*h*\nabla f.
\end{aligned}$$

Now we are able to compute the second variation of the Ricci tensor in the direction of h and k. By Lemma A.1, Lemma A.3 and Lemma A.4,

$$\begin{aligned}
\frac{d}{ds}\frac{d}{dt}|_{s,t=0}\mathrm{Ric}_{ij} =&\frac{d}{ds}\Big|_{s=0}\left(\frac{1}{2}\Delta_L h_{ij}-\delta^*(\delta h)_{ij}-\frac{1}{2}\nabla^2_{ij}(\mathrm{tr}h)\right)\\
=&\frac{1}{2}(\frac{d}{ds}\Big|_{s=0}\Delta_L)h_{ij}-(\frac{d}{ds}\Big|_{s=0}\delta^*)(\delta h)_{ij}-\delta^*(\frac{d}{ds}\Big|_{s=0}\delta h)_{ij}\\
&-\frac{1}{2}(\frac{d}{ds}\Big|_{s=0}\nabla^2_{ij})(\mathrm{tr}h)+\frac{1}{2}\nabla^2_{ij}\langle k,h\rangle\\
=&(k*\nabla^2 h)_{ij}+(\nabla^2 k*h)_{ij}+(\nabla k*\nabla h)_{ij}+(R*k*h)_{ij}.
\end{aligned}$$

For the scalar curvature, we therefore obtain, using Lemma A.2,

$$\begin{aligned}
\frac{d}{ds}\frac{d}{dt}|_{s,t=0}\mathrm{scal} &= \frac{d}{ds}\frac{d}{dt}\Big|_{s,t=0}(g^{ij}\mathrm{Ric}_{ij})\\
&=-k^{ij}\frac{d}{dt}\Big|_{t=0}\mathrm{Ric}_{ij}-h^{ij}\frac{d}{ds}\Big|_{s=0}\mathrm{Ric}_{ij}+g^{ij}\frac{d}{ds}\frac{d}{dt}\Big|_{s,t=0}\mathrm{Ric}_{ij}\\
&=k*\nabla^2 h+\nabla^2 k*h+\nabla k*\nabla h+R*k*h.
\end{aligned}$$

□

Index

$(.,.)_{L^2}$, L^2-scalar product, 6
$(E_i)_p$, fiber of the bundle E_i at p, 26
$(\delta_{g_E}^{-1}(0))^{C^{k,\alpha}}$, space of $C^{k,\alpha}$ divergence-free tensors , 88
$*$, Hamilton's notation, 69, 132
1-parameter family, 2
B, Bochner curvature tensor, 55
$C^\infty(M)$, space of smooth functions on M, 6
$C_g^\infty(M)$, 14
$C^{k,\alpha}(M)$, space of $C^{k,\alpha}$-functions on M, 97
$C_{g_E}^{k,\alpha}(M)$, 97
C_{g_E}, a map, 72
D, twisted Dirac operator, 22
D_1, a differential operator, 35
D_2, a differential operator, 35
E_6, an exceptional Lie group, 22
F_4, an exceptional Lie group, 22
G, Einstein tensor, 10
G, first variation of the Levi-Civita connection, 130
H, a differential operator, 60
$H^1(M)$, Sobolev space of functions on M, 45
H_1, space of hermitian tensors, 54
H_2, space of skew-hermitian tensors, 54
$H_{k,k}$, set of harmonic polynomials in $P_{k,k}$, 126
$Hol_p(M,g)$, Holonomy of (M,g) w.r.t. p, 25
J, almost complex structure, 54
K, Gaussian curvature, 9
K, sectional curvature, 36
K_{max} maximal sectional curvature, 36
K_{min}, minimal sectional curvature, 36
$L(V)$, space of linear maps on V, 26
$L^2(S^2M)$, space of L^2-sections of S^2M, 15
M^n, a manifold with dimension n, 5
O(n), orthogonal group, 22
O_1, an error term, 77, 108
O_2, an error term, 77, 108
$P_{k,k}$, set of homogeneous polnomials of degree k in z and $\bar{z}$, 126
R, Riemann curvature tensor, 5
S, Einstein-Hilbert functional, 9
S, spinor bundle, 22
SO(n), special orthogonal group, 22
SU(n), special unitary group, 22
S_g^2M, 35
S^n, sphere, 14
S^pM, bundle of symmetric $(0,p)$- tensors over M, 7
S_m, symmetric group, 51
Sc, scalar part of R, 51
Sp(n), symplectic group, 22
$Spin$(n), spin group, 22
TM, tangent bundle of M, 22
TT, transverse traceless tensors, 13
$T \circ S$, composition of symmetric $(0,2)$-tensors, 7
T^n, torus, 21
$T_g\mathcal{M}$, tangent space of $\mathcal{M}$ at g, 10
U, traceless Ricci part of R, 51
U(n), unitary group, 22
W, Weyl curvature tensor, 43
W^+, self-dual part of W, 50
W^-, anti-self-dual part of W, 50
W_{max}, 48
W_{min}, 48
$X(f)$, derivative of f along X, 7
$X^\flat$, flat of X, 6
Y, Yamabe functional, 19
$Y(M)$, Yamabe invariant of M, 19
$Y(M,[g])$, Yamabe constant of $[g]$, 19
$Y([g])$, Yamabe constant of $[g]$, 47
$[.,.]$, Lie-bracket of vector fields, 7
$[g]$, conformal class of g, 13
$\mathbb{C}P^n$, complex projective space, 17

Δ, Laplace-Beltrami operator, 7
Δ_0, Laplacian on functions, 21
Δ_1, connection Laplacian on $\Omega^1(M)$, 28
Δ_C, complex Laplacian, 55
Δ_E, Einstein operator, 15
Δ_H, Hodge Laplacian, 15, 55
Δ_L, Lichnerowicz Laplacian, 8
Γ, space of smooth sections of a vector bundle, 7
$\Gamma_g(S^2M)$, 11
Γ_{ij}^k, Christoffel symbol, 24
$\mathbb{H}P^n$, quaternionic projective space, 22
Λ^2M, bundle of 2-forms, 48
$\Omega^1(M)$, space of 1-forms on M, 6
par(M), space of parallel 1-forms on M, 30
$\mathbb{R}P^n$, real projective space, 22
$\|.\|_{C^k}$, C^k-norm, 6
$\|.\|_{C^{k,\alpha}}$, Hölder norm, 6
$\|.\|_{H^k}$, Sobolev norm of L^2-type, 6
$\|.\|_{L^p}$, L^p-norm, 6
$\|.\|_{W^{k,p}}$, Sobolev norm, 6
α_g, differential of $g \mapsto \Delta_g \mathrm{scal}_g$, 17
$\chi(M)$, Euler Characteristic of M, 9
δ, divergence, 7
δ^*, adjoint of δ, 7
$\delta^{-1}(0)$, space of divergence-free $(0,2)$-tensors, 13
δ_{ij}, Kronecker delta, 26
dim, dimension, 25
dV, volume element, 6
$⨍$, averaging integral, 64
grad$f$, gradient of the function f, 6
$\hat{R}$ Riemann curvature operator, 53
$\hat{W}$, Weyl curvature operator, 48
ind, index of a quadratic form, 32
ker, kernel of an operator, 16
$\lambda(g)$, Perelman's λ-functional, 59
$\langle .,. \rangle$, pointwise inner product, 6
$\mathcal{B}_\epsilon^k$, ϵ-ball w.r.t. the $C_{g_E}^k$-norm, 82
$\mathcal{C}$, set of metrics of constant scalar curvature, 17
$\mathcal{C}_c$, set of metrics of constant scalar curvature and volume c, 17
$\mathcal{E}$, set of Einstein metrics in a slice, 71
$\mathcal{L}_X$, Lie derivative along X, 7
$\mathcal{M}$, the set of smooth Riemannian metrics, 9
$\mathcal{M}^{C^{2,\alpha}}$, set of $C^{2,\alpha}$-metrics, 87
$\mathcal{M}_c$, set of smooth metrics with volume c, 11
$\mathcal{P}$, set of Einstein metrics with fixed constant in a slice, 71
$\mathcal{S}_{g_0}$, slice of the metric g_0, 16
$\mathcal{W}_+(g,f)$, 62
$\mathcal{W}_+(g,f,\sigma)$, 62
$\mathcal{W}_-(g,f,\tau)$, 94
$\mathcal{Y}_c$, Yamabe metrics of volume c, 19
$\mathfrak{X}(M)$, vector fields on M, 6
$\mathring{B}$, Bochner curvature action on S^2M, 56
$\mathring{R}$, curvature action on S^2M, 8
$\mathring{W}$, Weyl curvature action on S^2M, 43
End, endomorphism bundle, 25
dx, Euclidean volume element, 130
pr, projection map, 39
span, linear span, 42
$\mu_+(g)$, expander entropy, 63
$\mu_-(g,\tau)$, 94
mult$_\Delta(\lambda)$, multiplicity of λ as an eigenvalue of Δ, 32
∇, covariant derivative, 7
∇^k, k'th covariant derivative, 6
$\nu_-(g)$, shrinker entropy, 94
$\odot$, symmetric tensor product, 25
$\omega^\sharp$, sharp of ω, 6
$\otimes$, tensor product, 31
$\overline{w}(p)$, 48
$\owedge$, Kulkarni-Nomizu product, 43
∂_i, directional derivative, 130
∂_{x_i}, directional derivative, 126
$\rho(G)$, representation of G, 26
Ric, Ricci tensor, 5
Ric^0, traceless Ricci tensor, 43, 108
scal, scalar curvature, 5
spec, spectrum, 21
spec$_+$, positive spectrum, 18
τ_g, minimizer realizing $\nu_-(g)$, 94
Diff(M), group of diffeomorphism of M, 13
fd(M), flat dimension of M, 39
fd$(M)_p$, flat dimension of M at p, 39
tr, trace, 7
tr$^{-1}(0)$, space of traceless symmetric $(0,2)$-tensors, 13
$|.|$, pointwise norm, 6
$|.|_p$, pointwise norm at p, 48
vol(M,g), volume of (M,g), 11
$b(p)$, 57

Bibliography

[AM11] ANDERSSON, Lars ; MONCRIEF, Vincent: Einstein spaces as attractors for the Einstein flow. In: *J. Differ. Geom.* **89** (2011), no. 1, 1–47

[And05] ANDERSON, Michael T.: On uniqueness and differentiability in the space of Yamabe metrics. In: *Commun. Contemp. Math.* **7** (2005), no. 3, 299–310

[Böh98] BÖHM, Christoph: Inhomogeneous Einstein metrics on low-dimensional spheres and other low-dimensional spaces. In: *Invent. Math.* **134** (1998), no. 1, 145–176

[Böh05] BÖHM, Christoph: Unstable Einstein metrics. In: *Math. Z.* **250** (2005), no. 2, 279–286

[Bam10] BAMLER, Richard: Stability of symmetric spaces of noncompact type under Ricci flow. (2010). – arXiv:1011.4267

[Bam11] BAMLER, Richard: Stability of hyperbolic manifolds with cusps under Ricci flow. (2011). – arXiv:1004.2058v2

[Bar93] BARMETTLER, Urs: On the Lichnerowicz Laplacian. (1993). – PhD Thesis

[Bau09] BAUM, Helga: *Gauge theory. An introduction into differential geometry on fibre bundles. (Eichfeldtheorie. Eine Einführung in die Differentialgeometrie auf Faserbündeln.).* Berlin: Springer. xiv, 358 p., 2009

[Ber65] BERGER, Marcel: Sur les variétés d'Einstein compactes. In: *Comptes Rendus de la IIIe Réunion du Groupement des Mathématiciens d'Expression Latine.* 1965, 35–55

[Bes08] BESSE, Arthur L.: *Einstein manifolds. Reprint of the 1987 edition.* Berlin: Springer, 2008

[BGM71] BERGER, Marcel ; GAUDUCHON, Paul ; MAZET, Edmond: *Le spectre d'une variété riemannienne. (The spectrum of a Riemannian manifold).* Lecture Notes in Mathematics. 194. Berlin-Heidelberg-New York: Springer-Verlag. VII, 251 p., 1971

[Bie12] BIEBERBACH, Ludwig: Über die Bewegungsgruppen der euklidischen Räume. (Zweite Abhandlung.) Die Gruppen mit einem endlichen Fundamentalbereich. In: *Math. Ann.* **72** (1912), 400–412

[BM12] BHATTACHARYA, Atreyee ; MAITY, Soma: Some unstable critical metrics for the $L^{n/2}$-norm of the curvature tensor. (2012). – arXiv:1211.5774

[Bou99] BOUCETTA, Mohamed: Spectrum of the Lichnerowicz Laplacians on the spheres and the real projective spaces. (Spectre des Laplacien de Lichnerowicz sur les sphères et les projectifs réels.). In: *Publ. Mat., Barc.* **43** (1999), no. 2, 451–483

[Bre10] BRENDLE, Simon: *Ricci flow and the sphere theorem.* Graduate Studies in Mathematics 111. Providence, RI: American Mathematical Society (AMS). vii, 176 p., 2010

[BS09] BRENDLE, Simon ; SCHOEN, Richard: Manifolds with 1/4-pinched curvature are space forms. In: *J. Am. Math. Soc.* **22** (2009), no. 1, 287–307

[BWZ04] BÖHM, Christoph ; WANG, McKenzie Y. ; ZILLER, Wolfgang: A variational approach for compact homogeneous Einstein manifolds. In: *Geom. Funct. Anal.* 14 (2004), no. 4, 681–733

[Cao10] CAO, Huai-Dong: *Recent progress on Ricci solitons.* Lee, Yng-Ing (ed.) et al., Recent advances in geometric analysis. Proceeding of the international conference on geometric analysis, Taipei, Taiwan, June 18–22, 2007. Somerville, MA: International Press; Beijing: Higher Education Press. Advanced Lectures in Mathematics (ALM) 11, 1-38 (2010)., 2010

[CCG+07] CHOW, Bennett ; CHU, Sun-Chin ; GLICKENSTEIN, David ; GUENTHER, Christine ; ISENBERG, James ; IVEY, Tom ; KNOPF, Dan ; LU, Peng ; LUO, Feng ; NI, Lei: *The Ricci flow: techniques and applications. Part I: Geometric aspects.* Mathematical Surveys and Monographs 135. Providence, RI: American Mathematical Society (AMS). xxiii, 536 p., 2007

[CCG+08] CHOW, Bennett ; CHU, Sun-Chin ; GLICKENSTEIN, David ; GUENTHER, Christine ; ISENBERG, James ; IVEY, Tom ; KNOPF, Dan ; LU, Peng ; LUO, Feng ; NI, Lei: *The Ricci flow: techniques and applications. Part II: Analytic aspects.* Mathematical Surveys and Monographs 144. Providence, RI: American Mathematical Society (AMS). xxv, 458 p., 2008

[CDF84] CASTELLANI, Leonardo ; D'AURIA, Riccardo ; FRÉ, Pietro: SU(3)⊗SU(2)⊗U(1) from D = 11 supergravity. In: *Nuclear Phys.* 239 (1984), no. 2, 610–652

[CH13] CAO, Huai-Dong ; HE, Chenxu: Linear Stability of Perelmans ν-entropy on Symmetric spaces of compact type. (2013). – arXiv:1304.2697v1

[Cha86] CHARLAP, Leonard S.: *Bieberbach groups and flat manifolds.* New York, Springer, 1986

[CHI04] CAO, Huai-Song ; HAMILTON, Richard ; ILMANEN, Tom: Gaussian densities and stability for some Ricci solitons. (2004). – arXiv:math/0404165

[CK04] CHOW, Bennett ; KNOPF, Dan: *The Ricci flow: an introduction.* Providence, RI: American Mathematical Society (AMS), 2004

[CM12] COLDING, Tobias H. ; MINICOZZI, William P.: On uniqueness of tangent cones for Einstein manifolds. (2012). – arXiv:1206.4929

[CZ12] CAO, Huai-Dong ; ZHU, Meng: On second variation of Perelman's Ricci shrinker entropy. In: *Math. Ann.* **353** (2012), no. 3, 747–763

[Dai07] DAI, Xianzhe: Stability of Einstein Metrics and Spin Structures. In: *Proceedings of the 4th International Congress of Chinese Mathematicians* Vol II (2007), 59–72

[DFVN84] D'AURIA, Riccardo ; FRÉ, Pietro ; VAN NIEUWENHUIZEN, Peter: N = 2 matter coupled supergravity from compactification on a coset G/H possessing an additional Killing vector. In: *Phys. Lett.* **136** (1984), no. 5-6, 347–353

[Die13] DIETERICH, Peter-Simon: *On the Lichnerowicz Laplace operator and its application to stability of spacetimes*, Diplomarbeit, 2013

[DWW05] DAI, Xianzhe ; WANG, Xiaodong ; WEI, Guofang: On the stability of Riemannian manifold with parallel spinors. In: *Invent. Math.* **161** (2005), no. 1, 151–176

[DWW07] DAI, Xianzhe ; WANG, Xiaodong ; WEI, Guofang: On the variational stability of Kähler-Einstein metrics. In: *Commun. Anal. Geom.* **15** (2007), no. 4, 669–693

[Ebi70] EBIN, David G.: The manifold of Riemannian metrics. In: *Proc. Symp. AMS* Bd. 15, 1970, 11–40

[FH05] FLANAGAN, Eanna E. ; HUGHES, Scott A.: The basics of gravitational wave theory. In: *New Journal of Physics* **204** (2005), no. 7, 1–51

[FIN05] FELDMAN, Michael ; ILMANEN, Tom ; NI, Lei: Entropy and reduced distance for Ricci expanders. In: *J. Geom. Anal.* **15** (2005), no. 1, 49–62

[Fuj79] FUJITANI, Tamehiro: Compact suitable pinched Einstein manifolds. In: *Bull. Faculty Liberal Arts, Nagasaki Univ.* (1979), no. 19, 1–5

[GH02] GIBBONS, Gary W. ; HARTNOLL, Sean A.: Gravitational instability in higher dimensions. In: *Phys. Rev. D.* **66** (2002), no. 6

[GHP03] GIBBONS, Gary W. ; HARTNOLL, Sean A. ; POPE, Christopher N.: Bohm and Einstein-Sasaki Metrics, Black Holes, and Cosmological Event Horizons. In: *Phys. Rev. D* **67** (2003), no. 8

[GIK02] GUENTHER, Christine ; ISENBERG, James ; KNOPF, Dan: Stability of the Ricci flow at Ricci-flat metrics. In: *Commun. Anal. Geom.* **10** (2002), no. 4, 741–777

[GL99] GURSKY, Matthew J. ; LEBRUN, Claude: On Einstein manifolds of positive sectional curvature. In: *Ann. Global Anal. Geom.* **17** (1999), no. 4, 315–328

[GM02] GUBSER, Steven S. ; MITRA, Intrajit: Some interesting violations of the Breitenlohner-Freedman bound. In: *J. High Energy Phys* (2002), no. 7

[GPY82] GROSS, David J. ; PERRY, Malcolm J. ; YAFFE, Laurence G.: Instability of flat space at finite temperature. In: *Phys. Rev. D* **25** (1982), no. 2, 330–355

[Ham82] HAMILTON, Richard S.: Three-manifolds with positive Ricci curvature. In: *J. Differ. Geom.* **17** (1982), 255–306

[Ham95] HAMILTON, Richard S.: *The formation of singularities in the Ricci flow.* Hsiung, C. C. (ed.) et al., Proceedings of the conference on geometry and topology held at Harvard University, Cambridge, MA, USA, April 23-25, 1993. Cambridge, MA: International Press. Surv. Differ. Geom., Suppl. J. Differ. Geom. 2, 7-136., 1995

[Has12] HASLHOFER, Robert: Perelman's lambda-functional and the stability of Ricci-flat metrics. In: *Calc. Var. Partial Differ. Equ.* **45** (2012), no. 3-4, 481–504

[Hil15] HILBERT, David: Die Grundlagen der Physik. (Erste Mitteilung.). In: *Gött. Nachr.* (1915), 395–407

[HM13] HASLHOFER, Robert ; MÜLLER, Reto: Dynamical stability and instability of Ricci-flat metrics. (2013). – arXiv:1301.3219

[Hui85] HUISKEN, Gerhard: Ricci deformation of the metric on a Riemannian manifold. In: *J. Differ. Geom.* **21** (1985), 47–62

[IK04] ITOH, Mitsuhiro ; KOBAYASHI, Daisuke: Isolation theorems of the Bochner curvature type tensors. In: *Tokyo J. Math.* **27** (2004), no. 1, 227–237

[IN05] ITOH, Mitsuhiro ; NAKAGAWA, Tomomi: Variational stability and local rigidity of Einstein metrics. In: *Yokohama Math. J.* **51** (2005), no. 2, 103–115

[IS02] ITOH, Mitsuhiro ; SATOH, Hiroyasu: Isolation of the Weyl conformal tensor for Einstein manifolds. In: *Proc. Japan Acad., Ser. A* **78** (2002), no. 7, 140–142

[Kan06] KANG, Eun S.: Moduli spaces of 3-dimensional flat manifolds. In: *J. Korean Math. Soc.* **43** (2006), no. 5, 1065–1080

[KK03] KANG, Eun S. ; KIM, Ju Y.: Deformation spaces of 3-dimensional flat manifolds. In: *Commun. Korean Math. Soc.* **18** (2003), no. 1, 95–104

[Koi78] KOISO, Norihito: Non-deformability of Einstein metrics. In: *Osaka J. Math.* **15** (1978), 419–433

[Koi79a] KOISO, Norihito: A decomposition of the space $\mathcal{M}$ of Riemannian metrics on a manifold. In: *Osaka J. Math.* **16** (1979), 423–429

[Koi79b] KOISO, Norihito: On the second derivative of the total scalar curvature. In: *Osaka J. Math.* **16** (1979), 413–421

[Koi80] KOISO, Norihito: Rigidity and stability of Einstein metrics - The case of compact symmetric spaces. In: *Osaka J. Math.* **17** (1980), 51–73

[Koi82] KOISO, Norihito: Rigidity and infinitesimal deformability of Einstein metrics. In: *Osaka J. Math.* **19** (1982), 643–668

[Koi83] KOISO, Norihito: Einstein metrics and complex structures. In: *Invent. Math.* **73** (1983), 71–106

[KW75] KAZDAN, Jerry L. ; WARNER, Frank W.: *Prescribing curvatures.* Differ. Geom., Proc. Symp. Pure Math. 27, Part 2, Stanford, 1975

[LeB99] LEBRUN, Claude: *Einstein metrics and the Yamabe problem.* Alexiades, Vasilios (ed.) et al., Trends in mathematical physics. Proceedings of the conference, University of Tennessee, Knoxville, TN, USA, October 14–17, 1998. Providence, RI: American Mathematical Society. AMS/IP Stud. Adv. Math. 13, 353-376., 1999

[Lic61] LICHNEROWICZ, André: Propagateurs et commutateurs en relativité générale. In: *Publications Mathématiques de l'IHÉS* **10** (1961), no. 1, 5–56

[LP87] LEE, John M. ; PARKER, Thomas H.: The Yamabe problem. In: *Bull. Am. Math. Soc., New Ser.* **17** (1987), 37–91

[Mut69] MUTO, Yosio: Einstein spaces of positive scalar curvature. In: *J. Differ. Geom.* **3** (1969), 457–459

[NS82] NAIMARK, Mark A. ; STERN, Aleksandr I.: *Theory of group representations. Transl. from the Russian by Elizabeth Hewitt, ed. by Edwin Hewitt.* , 1982

[Oba62] OBATA, Morio: Certain conditions for a Riemannian manifold to be isometric with a sphere. In: *J. Math. Soc. Japan* **14** (1962), 333–340

[Omo68] OMORI, Hideki: On the group of diffeomorphisms on a compact manifold. **15** (1968), 167–183

[Per02] PERELMAN, Grisha: The entropy formula for the Ricci flow and its geometric applications. (2002). – arXiv:math/0211159

[Per03] PERELMAN, Grisha: Ricci flow with surgery on three-manifolds. (2003). – arXiv:math/0303109

[PP84a] PAGE, Don N. ; POPE, Christopher N.: Stability analysis of compactifications of D = 11 supergravity with SU(3)×SU(2)×U(1) symmetry. In: *Phys. Lett.* **145** (1984), no. 5, 337–341

[PP84b] PAGE, Don N. ; POPE, Christopher N.: Which compactifications of D = 11 supergravity are stable? In: *Phys. Lett.* **144** (1984), no. 5, 346–350

[Rom85] ROMANS, Larry J.: New compactifications of chiral N = 2, d = 10 supergravity. In: *Phys. Lett.* **153** (1985), no. 6, 392–396

[Sak71] SAKAI, Takashi: On eigen-values of Laplacian and curvature of Riemannian manifold. In: *Tohoku Math. J., II. Ser.* **23** (1971), 589–603

[Sch84] SCHOEN, Richard M.: Conformal deformation of a Riemannian metric to constant scalar curvature. In: *J. Differ. Geom.* **20** (1984), 479–495

[Sch89] SCHOEN, Richard M.: *Variational theory for the total scalar curvature functional for Riemannian metrics and related topics.* Topics in calculus of variations, Lect. 2nd Sess., Montecatini/Italy 1987, Lect. Notes Math. 1365, 120-154., 1989

[Ses06] SESUM, Natasa: Linear and dynamical stability of Ricci-flat metrics. In: *Duke Math. J.* **133** (2006), no. 1, 1–26

[Sin92] SINGER, Michael: Positive Einstein metrics with small $L^{n/2}$-norm of the Weyl tensor. In: *Differ. Geom. Appl.* **2** (1992), no. 3, 269–274

[Smi75] SMITH, Robert T.: The second variation formula for harmonic mappings. In: *Proc. Am. Math. Soc.* **47** (1975), 229–236

[SSS11] SCHNÜRER, Oliver C. ; SCHULZE, Felix ; SIMON, Miles: Stability of hyperbolic space under Ricci flow. In: *Commun. Anal. Geom.* **19** (2011), no. 5, 1023–1047

[SW13] SUN, Song ; WANG, Yuanqi: On the Kähler-Ricci flow near a Kähler-Einstein metric. (2013). – arXiv:1004.2018v3

[Wan91] WANG, McKenzie Y.: Preserving parallel spinors under metric deformations. In: *Indiana Univ. Math. J.* **40** (1991), no. 3, 815–844

[Wol11] WOLF, Joseph A.: *Spaces of constant curvature. 6th ed.* Providence, RI: AMS Chelsea Publishing. xv, 420 p., 2011

[WZ86] WANG, McKenzie ; ZILLER, Wolfgang: Einstein metrics with positive scalar curvature. In: *Curvature and topology of Riemannian manifolds (Katata, 1985).* Lecture Notes in Math., 1201, Springer, Berlin, 1986, 319–336

[Ye93] YE, Rugang: Ricci flow, Einstein metrics and space forms. In: *Trans. Am. Math. Soc.* **338** (1993), no. 2, 871–896

[Zhu00] ZHU, Chenchang: The Gauss-Bonnet Theorem and its Applications. In: *University of California, Berkeley, USA* (2000)

MIX
Papier aus verantwortungsvollen Quellen
Paper from responsible sources
FSC® C105338

Printed by Books on Demand GmbH, Norderstedt / Germany